This book is a major new contribution to decision theory, focusing on the question of when it is rational to accept scientific theories.

The author examines both Bayesian decision theory and confirmation theory, refining and elaborating the views of Ramsey and Savage. He argues that the most solid foundations for confirmation theory are to be found in decision theory, and he provides a decision-theoretic derivation of principles for how new probabilities should be revised over time. Professor Maher defines a notion of accepting a hypothesis, and then shows that it is not reducible to probability and that it is needed to deal with some important questions in the philosophy of science. A Bayesian decision-theoretic account of rational acceptance is provided, together with a proof of the foundations for this theory. A final chapter shows how this account can be used to cast light on such vexed issues as verisimilitude and scientific realism.

This is a book of critical importance to all philosophers of science and epistemologists, as well as to decision theorists in economics and other branches of the social sciences.

T0275757

Betting on theories

Cambridge Studies in Probability, Induction, and Decision Theory

General editor: Brian Skyrms

Advisory editors: Ernest W. Adams, Ken Binmore, Jeremy Butterfield, Persi Diaconis, William L. Harper, John Harsanyi, Richard C. Jeffrey, Wolfgang Spohn, Patrick Suppes, Amos Tversky, Sandy Zabell

This new series is intended to be the forum for the most innovative and challenging work in the theory of rational decision. It focuses on contemporary developments at the interface between philosophy, psychology, economics, and statistics. The series addresses foundational theoretical issues, often quite technical ones, and therefore assumes a distinctly philosophical character.

Other titles in the series
Ellery Eells, *Probabilistic Causality*
Richard Jeffrey, *Probability and the Art of Judgment*
Robert C. Koons, *Paradoxes of Belief and Strategic Rationality*
Cristina Bicchieri and Maria Luisa Dalla Chiara (eds.), *Knowledge, Belief, and Strategic Interaction*

Forthcoming
J. Howard Sobel, *Taking Chances*
Patrick Suppes and Mario Zanotti, *Foundations of Probability with Applications*
Clark Glymour and Kevin Kelly (eds.), *Logic, Computation, and Discovery*

Betting on theories

Patrick Maher

University of Illinois at Urbana-Champaign

Published by the Press Syndicate of the University of Cambridge
The Pitt Building, Trumpington Street, Cambridge CB2 1RP
40 West 20th Street, New York, NY 10011-4211, USA
10 Stamford Road, Oakleigh, Victoria 3166, Australia

© Cambridge University Press 1993

First published 1993

Library of Congress Cataloging-in-Publication Data
Maher, Patrick.
Betting on theories / Patrick Maher.
p.   cm. – (Cambridge studies in probability, induction, and
decision theory)
Includes bibliographical references and index.
ISBN 0-521-41850-X
1. Decision-making.   I. Title.   II. Series.
QA279.4.M33   1993
519.5'42 – dc20   92-13817
CIP

A catalog record for this book is available from the British Library

ISBN 0-521-41850-X

Transferred to digital printing 2004

Contents

Preface

Under what conditions does evidence confirm a scientific hypothesis? And why under those conditions only? There is an answer to these questions that is both precise and general, and which fits well with scientific practice. I allude to the Bayesian theory of confirmation. This theory represents scientists as having subjective probabilities for hypotheses, and it uses probability theory (notably Bayes' theorem) to explain when, and why, evidence confirms scientific theories.

I think Bayesian confirmation theory is correct as far as it goes and represents a great advance in the theory of confirmation. But its foundations have sometimes been seen as shaky. Can we really say that scientists have subjective probabilities for scientific theories – or even that rationality requires this? One purpose of the present book is to address this foundational issue. In Chapter 1, I defend an interpretation of subjective probability that is in the spirit of Frank Ramsey. On this interpretation, a person has subjective probabilities if the person has preferences satisfying certain conditions. In Chapters 2 and 3 I give reasons for thinking that these conditions are requirements of rationality. It follows that rational people have (not necessarily precise) subjective probabilities. In Chapter 4, I apply this general argument to science, to conclude that rational scientists have (not necessarily precise) subjective probabilities for scientific theories.

The presupposition of Bayesian confirmation theory that I have been discussing is a *synchronic* principle of rationality; it says that a rational scientist at a given time has subjective probabilities. But Bayesian confirmation theory also assumes principles about how these probabilities should be revised as new information is acquired; these are *diachronic* principles of rationality. Such principles are investigated in Chapter 5. Here

I show that the usual arguments for these diachronic principles are fallacious. Furthermore, if the arguments were right, they would legitimate these principles in contexts where they are really indefensible. I offer a more modest, and I hope more cogent, replacement for these flawed arguments.

Non-Bayesian philosophers of science have focused on the question of when it is rational to *accept* a scientific theory. Bayesians have tended to regard their notion of subjective probability as a replacement for the notion of acceptance and thus to regard talk of acceptance as a loose way of talking about subjective probabilities. Insofar as the notion of acceptance has been poorly articulated in non-Bayesian philosophy of science, this cavalier attitude is understandable. But there is a notion of acceptance that can be made clear and is not reducible to the concept of subjective probability. Confirmation theory does not provide a theory of rational acceptance in this sense. Instead, we need to employ decision theory. This is argued in Chapter 6.

In my experience, Bayesian philosophers of science tend to think that if acceptance is not reducible to subjective probability then it cannot be an important concept for philosophy of science. But this thought is mistaken. One reason is that Bayesian analyses of the history of science, which are used to argue for the correctness of Bayesian confirmation theory, themselves require a theory of acceptance. Another reason is the acknowledged role of alternative hypotheses in scientific development; accounting for this role is beyond the resources of confirmation theory and requires the theory of acceptance. A third reason is that we would like to explain why gathering evidence contributes to scientific goals, and acceptance theory provides a better account of this than is possible using confirmation theory alone. These claims are defended in Chapter 7.

A decision-theoretic account of rational acceptance raises foundational questions again. We now require not only that scientists have probabilities for scientific hypotheses but also that acceptance of hypotheses have consequences to which utilities can be assigned. Since acceptance is a cognitive act, the consequences in question can be called cognitive consequences, and their utilities can be called cognitive utilities. It needs to be shown that the required measures of cognitive utility can meaningfully be assigned. According to the approach defended in Chapter 1, this requires showing that rational scientists have preferences among cognitive options that can be represented by a pair of probability and cognitive utility functions. Chapter 8 takes up this challenge, and answers it by providing a new representation theorem.

In the final chapter of the book, I try to show that my theory of acceptance provides a fruitful perspective on traditional questions about scientific values. Here I partially side with those who see science as aiming at truth. In fact, I show how it is possible to define a notion of verisimilitude (or closeness to the whole truth), and I argue that this is ultimately the only scientific value. But on the other hand, I differ from most realists in allowing that different scientists may, within limits, assess verisimilitude in different ways. The upshot is a position on scientific values that is sufficiently broad-minded to be consistent with van Fraassen's antirealism, but is strong enough to be inconsistent with Kuhn's conception of scientific values.

I began work on the acceptance theory in this book in the winter of 1980, when I was a graduate student at the University of Pittsburgh. That semester, seminars by both Carl Hempel and Teddy Seidenfeld introduced me to Hempel's and Levi's decision-theoretic models of rational acceptance. I later wrote my dissertation (Maher 1984) on this topic, proving the representation theorem described in Chapter 8 and developing the subjective theory of verisimilitude presented in Chapter 9. Wesley Salmon gently directed my dissertation research, other members of my dissertation committee (David Gauthier, Clark Glymour, Carl Hempel, and Nicholas Rescher) provided valuable feedback, and Teddy Seidenfeld helped enormously, especially on the representation theorem.

In 1987, I received a three-year fellowship from the Michigan Society of Fellows, a grant from the National Science Foundation, a summer stipend from the National Endowment for the Humanities, and a humanities released-time award from the University of Illinois – all to write this book. Eventually I decided that the focus of the book needed to be narrowed; hence much of the research supported by these awards has been excluded from this book, although most has been published elsewhere. Still, without those awards, especially the Michigan fellowship, this book would never have been written.

I gave seminars based on drafts of this book at Michigan and Illinois; I thank the participants in those seminars for discussions that helped improve the book. Critiques by George Mavrodes and Wei-ming Wu are especially vivid in my mind. Brad Armendt read the book for Cambridge University Press, and his written comments were very useful to me. When I thought the book was basically finished, Allan Gibbard's (1990) book prompted me to change radically the conception of rationality I was working with; I thank my Illinois colleagues who read through Gibbard's book with me, and thank Gibbard for an illuminating e-mail correspondence. Fred Schmitt and Steven Wagner gave me stimulating written comments on the material in Chapter 6. Many others discussed aspects of the book with me, and I hope they will forgive me if I don't drag these acknowledgments out to an interminable length by attempting to list them all. I will, however, mention my wife, Janette, who not only provided domestic support but also read much of the book, talked about it with me on and off for years, and suggested improvements. I will also mention my trusty PCs and Leslie Lamport's (1986) LaTeX software, which together turned countless drafts into beautiful pages. LaTeX deserves to be more widely used by philosophers than it is, especially since it is free.

Chapter 5 is an improved version of my "Diachronic Rationality," *Philosophy of Science* **59** (1992). Section 7.3 is a refined version of an argument first published in "Why Scientists Gather Evidence," *British Journal for the Philosophy of Science* **41** (1990). Section 4.3, and parts of Chapter 6 and Section 9.6, appeared in "Acceptance Without Belief," *PSA 1990*, vol. 1. This material is reproduced here by permission of the publishers.

1

The logic of preference

The heart of Bayesian theory is the principle that rational choices maximize expected utility. This chapter begins with a statement of that principle. The principle is a formal one, and what it means is open to some interpretation. The remainder of the chapter is concerned with setting out an interpretation that makes the principle both correct and useful. I also indicate how I would defend these claims of correctness and usefulness.

1.1 EXPECTED UTILITY

If you need to make a decision, then there is more than one possible *act* that you could choose. In general, these acts will have different *consequences*, depending on what the true *state* of the world may be; and typically one is not certain which state that is. *Bayesian decision theory* is a theory about what counts as a rational choice in a decision problem. The theory postulates that a rational person has a probability function p defined over the states, and a utility function u defined over the consequences. Let $a(x)$ denote the consequence that will be obtained if act a is chosen and state x obtains, and let X be the set of all possible states. Then the *expected utility* of act a is the expected value of $u(a(x))$; I will refer to it as $EU(a)$. If X is countable, we can write

$$EU(a) = \sum_{x \in X} p(x)u(a(x)).$$

Bayesian decision theory holds that the choice of act a is rational just in case the expected utility of a is at least as great as that of any other available act. That is, rational choices maximize expected utility.

The principle of maximizing expected utility presupposes that the acts, consequences, and states have been formulated

1

appropriately. The formulation is appropriate if the decision maker is (or ought to be) sure that

1. one and only one state obtains;
2. the choice of an act has no causal influence on which state obtains;[1] and
3. the consequences are sufficiently specific that they determine everything that is of value in the situation.

The following examples illustrate why conditions 2 and 3 are needed.

Mr. Coffin is a smoker considering whether to quit or continue smoking. All he cares about is whether or not he smokes and whether or not he lives to age 65, so he takes the consequences to be

Smoke and live to age 65
Quit and live to age 65
Smoke and die before age 65
Quit and die before age 65

The first-listed consequence has highest utility for Coffin, the second-listed consequence has second-highest utility, and so on down. And Coffin takes the states to be "Live to age 65" and "Die before age 65." Then each act–state pair determines a unique consequence, as in Figure 1.1. Applying the principle of maximizing expected utility, Coffin now reaches the conclusion that smoking is the rational choice. For he sees that whatever state obtains, the consequence obtained from smoking has higher utility than that obtained from not smoking; and so the expected utility of smoking is higher than that of not smoking. But if Coffin thinks that smoking might reduce the chance of living to age 65, then his reasoning is clearly faulty, for he has not taken account of this obviously relevant possibility. The fault lies in using states ("live to 65," "die before 65") that may be causally influenced by what is chosen, in violation of condition 2.

---

[1]Richard Jeffrey (1965, 1st ed.) maintained that what was needed was that the states be *probabilistically* independent of the acts. For a demonstration that this is not the same as requiring causal independence, and an argument that causal independence is in fact the correct requirement, see (Gibbard and Harper 1978).

	Live to 65	Die before 65
Smoke	Smoke and live to 65	Smoke and die before 65
Quit	Quit and live to 65	Quit and die before 65

Figure 1.1: Coffin's representation of his decision problem

Suppose Coffin is sure that the decision to smoke or not has no influence on the truth of the following propositions:

A: If I continue smoking then I will live to age 65.
B: If I quit smoking then I will live to age 65.

Then condition 2 would be satisfied by taking the states to be the four Boolean combinations of A and B (i.e., "A and B," "A and not B," "B and not A," and "neither A nor B"). Also, these states uniquely determine what consequence will be obtained from each act. And with these states, the principle of maximizing expected utility no longer implies that the rational choice is to smoke; the rational choice will depend on the probabilities of the states and the utilities of the consequences.

Next example: Ms. Drysdale is about to go outside and is wondering whether to take an umbrella. She takes the available acts to be "take umbrella" and "go without umbrella," and she takes the states be "rain" and "no rain." She notes that with these identifications, she has satisfied the requirement of act–state independence. Finally, she identifies the consequences as being that she is "dry" or "wet." So she draws up the matrix shown in Figure 1.2. Because she gives higher utility to staying dry than getting wet, she infers that the expected utility of taking the umbrella is higher than that of going without it, provided only that her probability for rain is not zero. Drysdale figures that the probability of rain is never zero, and takes her umbrella.

Since a nonzero chance of rain is not enough reason to carry an umbrella, Drysdale's reasoning is clearly faulty. The trouble

3

	Rain	No rain
Take umbrella	Dry	Dry
Go without	Wet	Dry

Figure 1.2: Drysdale's representation of her decision problem

	Rain	No rain
Take umbrella	Dry & umbrella	Dry & umbrella
Go without	Wet & no umbrella	Dry & no umbrella

Figure 1.3: Corrected representation of Drysdale's decision problem

is that carrying the umbrella has its own disutility, which has not been included in the specification of the consequences; this violates condition 3. If we include in the consequences a specification of whether or not the umbrella is carried, the consequences become those shown in Figure 1.3.

Suppose that these consequences are ranked by utility in this order:

Dry & no umbrella
Dry & umbrella
Wet & no umbrella

A mere positive probability for rain is now not enough to make taking the umbrella maximize expected utility; a small risk of getting wet would be worth running, for the sake of not having to carry the umbrella.

It is implicit in the definition of expected utility that each act has a unique consequence in any given state. This together with the previous conditions prevents the principle of maximizing expected utility being applied to cases where the laws of nature and the prior history of the world, together with the act chosen, do not determine everything of value in the situation (as might happen when the relevant laws are quantum mechanical).

4

In such a situation, taking the states to consist of the laws of nature and prior history of the world (or some part thereof) would not give a unique consequence for each state, unless the consequences omitted something of value in the situation. Including in the states a specification of what consequence will in fact be obtained avoids this problem but violates the requirement that the states be causally independent of the acts. There is a generalization of the principle of maximizing expected utility that can deal with decision problems of this kind; but I shall not present it here, because it introduces complexities that are irrelevant to the themes of this book. The interested reader is referred to (Lewis 1981).

1.2 CALCULATION

In many cases, it would not be rational to bother doing a calculation to determine which option maximizes expected utility. So if Bayesian decision theory held that a rational person would always do such calculations, the theory would be obviously incorrect. But the theory does not imply this.

To see that the theory has no such implication, note that doing a calculation to determine what act maximizes expected utility is itself an act; and this act need not maximize expected utility. For an illustration, consider again the problem of whether to take an umbrella. A fuller representation of the acts available would be the following:

t: Take umbrella, without calculating expected utility.

$\bar{t}$: Go without umbrella, without calculating expected utility.

c: Calculate the expected utility of t and $\bar{t}$ (with a view to subsequently making a choice that is calculated to maximize expected utility).[2]

Because calculation takes time, it may well be that t or $\bar{t}$ has higher expected utility than c; and if so, then Bayesian decision theory itself endorses not calculating expected utility.

[2] After calculating expected utility, one would choose an act without again calculating expected utility; thus the choice at that time will be between t and $\bar{t}$. So whether or not expected utility is calculated, one eventually chooses t or $\bar{t}$.

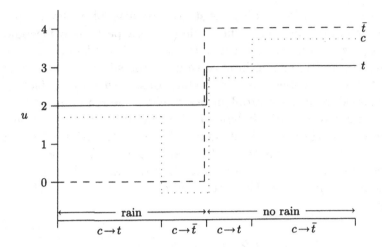

Figure 1.4: Acts of taking an umbrella (t), not taking umbrella ($\bar{t}$), and calculating (c) whether to choose t or $\bar{t}$.

Conversely, if c has higher expected utility than t or $\bar{t}$, then the theory holds that it is rational to do the calculation.

If we wanted to, we could do an expected utility calculation, to determine which of t, $\bar{t}$, and c maximizes expected utility. The situation is represented in Figure 1.4. Here "$c \to t$" means that choosing c (calculating the expected utility of t and $\bar{t}$) would lead to t being chosen;[3] and similarly for "$c \to \bar{t}$." When $c \to t$ is true, the utility of c is equal to that of t, less the cost of calculation; and when $c \to \bar{t}$ is true, the utility of c is equal to that of $\bar{t}$, less the same cost of calculation. Suppose that[4]

$$p(\text{rain}.c \to t) = p(\text{no rain}.c \to \bar{t}) \quad = \quad .4$$
$$p(\text{rain}.c \to \bar{t}) = p(\text{no rain}.c \to t) \quad = \quad .1$$
$$u(\bar{t}.\text{rain}) = 0; \quad u(t.\text{no rain}) \quad = \quad 3$$
$$u(t.\text{rain}) = 2; \quad u(\bar{t}.\text{no rain}) \quad = \quad 4$$

---

[3] Assuming one will choose an act that is calculated to maximize expected utility, $c \to t$ includes all states in which calculation would show t to have a higher expected utility than $\bar{t}$. But it may also include states in which calculation would show t and $\bar{t}$ to have the same expected utility.

[4] Here the dot represents conjunction, and its scope extends to the end of the formula. For example, $p(\text{no rain}.c \to t)$ is the probability that there there is no rain and that $c \to t$.

Then
$$EU(t) = 2.5; \qquad EU(\bar{t}) = 2$$
and letting x be the cost of calculation,
$$EU(c) = 2.7 - x.$$

Thus Bayesian decision theory deems c the rational choice if x is less than 0.2, but t is the rational choice if x exceeds 0.2.

In the course of doing this second-order expected utility calculation, I have also done the first-order calculation, showing that $EU(t) > EU(\bar{t})$. But this does not negate the point I am making, namely that Bayesian decision theory can deem it irrational to calculate expected utility. For Bayesian decision theory also does not require the second-order calculation to be done. This point will be clear if we suppose that you are the person who has to decide whether or not to take an umbrella, and I am the one doing the second-order calculation of whether you should do a first-order calculation. Then I can calculate (using your probabilities and utilities) that you would be rational to choose t, and irrational to choose c; and this does not require you to do any calculation at all. Likewise, I could if I wished (and if it were true) show that you would be irrational to do the second-order calculation which shows that you would be irrational to do the first-order calculation.[5]

This conclusion may at first sight appear counterintuitive. For instance, suppose that t maximizes expected utility and, in particular, has higher expected utility than both $\bar{t}$ and c. Suppose further that you currently prefer $\bar{t}$ to the other options and would choose it if you do not do any calculation. Thus if you do no calculation, you will make a choice that Bayesian decision theory deems irrational. But if you do a calculation, you also do something deemed irrational, for calculation has a lower expected utility than choosing t outright. This may seem

[5]Kukla (1991) discusses the question of when reasoning is rational, and sees just these options: (a) reasoning is rationally required only when we *know* that the benefits outweigh the costs; or (b) a metacalculation of whether the benefits outweigh the costs is always required. Since (b) is untenable, he opts for (a). But he fails to consider the only option consistent with decision theory and the one being advanced here: That reasoning and metacalculations alike are rationally required just when the benefits *do* outweigh the costs, whether or not it is known that they do.

7

to put you in an impossible position. To know that t is the rational choice you would need to do a calculation, but doing that calculation is itself irrational. You are damned if you do and damned if you don't calculate.

This much is right: In the case described, what you would do if you do not calculate is irrational, and so is calculating. But this does not mean that decision theory deems you irrational no matter what you do. In fact, there is an option available to you that decision theory deems rational, namely t. So there is no violation here of the principle that 'ought' implies 'can'.

What the case shows is that Bayesian decision theory does not provide a means of guaranteeing that your choices are rational. I suggest that expecting a theory of rational choice to do this is expecting too much. What we can reasonably ask of such a theory is that it provide a criterion for when choices are rational, which can be applied to actual cases, even though it may not be rational to do so; Bayesian decision theory does this.[6]

Before leaving this topic, let me note that in reality we have more than the two options of calculating expected utility and choosing without any calculation. One other option is to calculate expected utility for a simplified representation that leaves out some complicating features of the real problem. For example, in a real problem of deciding whether or not to take an umbrella we would be concerned, not just with whether or not it rains, but also with how much rain there is and when it occurs; but we could elect to ignore these aspects and do a calculation using the simple matrix I have been using here. This will maximize expected utility if the simplifications reduce the computational costs sufficiently without having too great a probability of leading to the wrong choice. Alternatively, it might maximize expected utility to use some non-Bayesian rule, such as minimizing the maximum loss or settling for an act in which all the outcomes are "satisfactory."[7]

---

[6]Railton (1984) argues for a parallel thesis in ethics. Specifically, he contends that morality does not require us to always calculate the ethical value of acts we perform, and may even forbid such calculation in some cases.

[7]This is the Bayes/non-Bayes compromise advocated by I. J. Good (1983, 1988), but contrary to what Good sometimes says, the rationale for the compromise does not depend on probabilities being indeterminate.

8

1.3 REPRESENTATION

Bayesian decision theory postulates that rational persons have the probability and utility functions needed to define expected utility. What does this mean, and why should we believe it?

I suggest that we understand attributions of probability and utility as essentially a device for interpreting a person's preferences. On this view, an attribution of probabilities and utilities is correct just in case it is part of an overall interpretation of the person's preferences that makes sufficiently good sense of them and better sense than any competing interpretation does. This is not the place to attempt to specify all the criteria that go into evaluating interpretations, nor shall I attempt to specify how good an interpretation must be to be sufficiently good. For present purposes, it will suffice to assert that if a person's preferences all maximize expected utility relative to some p and u, then it provides a perfect interpretation of the person's preferences to say that p and u are the person's probability and utility functions. Thus, having preferences that all maximize expected utility relative to p and u is a sufficient (but not necessary) condition for p and u to be one's probability and utility functions. I shall call this the *preference interpretation* of probability and utility.[8] Note that on this interpretation, a person can have probabilities and utilities without consciously assigning any numerical values as probabilities or utilities; indeed, the person need not even have the concepts of probability and utility.

Thus we can show that rational persons have probability and utility functions if we can show that rational persons have preferences that maximize expected utility relative to some such functions. An argument to this effect is provided by *representation theorems* for Bayesian decision theory. These theorems show that if a person's preferences satisfy certain putatively reasonable qualitative conditions, then those preferences are indeed representable as maximizing expected utility relative to some probability and utility functions. Ramsey (1926) and Savage (1954) each proved a representation theorem, and there have

---

[8]The preference interpretation is (at least) broadly in agreement with work in philosophy of mind, e.g., by Davidson (1984, pp. 159f.).

been many subsequent theorems, each making somewhat different assumptions. (For a survey of representation theorems, see [Fishburn 1981].)

As an illustration, and also to prepare the way for later discussion, I will describe two of the central assumptions used in Savage's (1954) representation theorem. First, we need to introduce the notion of *weak preference*. We say that you weakly prefer g to f if you either prefer g to f, or else are indifferent between them. The notation '$f \precsim g$' will be used to denote that g is weakly preferred to f. Now Savage's first postulate can be stated: It is that for any acts f, g, and h, the following conditions are satisfied:

Connectedness. *Either $f \precsim g$ or $g \precsim f$ (or both).*

Transitivity. *If $f \precsim g$ and $g \precsim h$, then $f \precsim h$.*

A relation that satisfies both the conditions of connectedness and transitivity is said to be a *weak* (or simple) *order*. So an alternative statement of this postulate is that the relation $\precsim$ is a weak order on the set of acts.

Savage's second postulate asserts that if two acts have the same consequences in some states, then the person's preferences regarding those acts should be independent of what that common consequence is. For example, in Figure 1.5, f and g have the same consequence on $\bar{A}$, and f' and g' are the result of replacing that common consequence with something else; so according to this postulate, if $f \precsim g$, then it should be that $f' \precsim g'$. Formally, the postulate is that for any acts f, g, f', and g', and for any event A, the following condition holds:[9]

Independence. *If $f = f'$ on A, $g = g'$ on A, $f = g$ on $\bar{A}$, $f' = g'$ on $\bar{A}$, and $f \precsim g$, then $f' \precsim g'$.*

---

[9]This postulate is often referred to as the *sure-thing principle*, a term that comes from Savage (1954, p. 21). But as I read Savage, what he means by the sure-thing principle is not any postulate of his theory, but rather an informal principle that motivates the present postulate. In Section 3.2.3 I will discuss that principle, and consider how well it motivates the postulate. So for my purposes, it would confuse an important distinction to refer to this postulate as "the sure-thing principle."

10

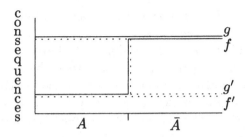

Figure 1.5: Illustration of the independence postulate

(The notation '$f = g$ on A' means that f and g have the same consequence for every state in A, i.e., $f(x) = g(x)$ for all $x \in A$.)

Savage shows that if your preferences satisfy these and other assumptions, then they maximize expected utility relative to some probability and utility functions. But it is worth noting that for transitivity and independence, the converse is also true; that is, if your preferences maximize expected utility relative to some probability and utility functions, then transitivity and independence must hold.

Proof. (Proofs that are set off from the text like this may be skipped without loss of continuity.) Suppose preferences maximize expected utility relative to some probability and utility functions. Then if $f \precsim g$ and $g \precsim h$, we have that $EU(f) \leq EU(g)$ and $EU(g) \leq EU(h)$, whence $EU(f) \leq EU(h)$, and $f \precsim h$; thus transitivity is satisfied. Now suppose the assumptions of the independence axiom are satisfied. It is straightforward to show that

$$EU(g) - EU(f) = EU(g') - EU(f').$$

Since $f \precsim g$, $EU(f) \leq EU(g)$, which by the above identity implies that $EU(f') \leq EU(g')$, and hence that $f' \precsim g'$, as the independence axiom requires.

Thus *any* representation theorem must either assume transitivity and independence or else assume something at least as strong as these conditions.

11

I do not hold that rationality *always* forbids violations of transitivity or independence. For example, an anti-Bayesian tycoon might offer me a million dollars to have preferences that are intransitive in some insignificant way; then I would agree that if I could make my preferences intransitive, this would be the rational thing for me to do. More realistically, it may be that my preferences are intransitive, but it would take more effort than it is worth to remove the intransitivities, in which case the rational thing to do would be to remain intransitive.

What I do hold is that when the preferences are relevant to a sufficiently important decision problem, and where there are no rewards attached to violating transitivity or independence, then it is rational to have one's preferences satisfy these conditions. In Chapters 2 and 3, I offer arguments to support this position, and I will critique arguments in the literature that, if sound, would refute this position.

It is too cumbersome to keep saying that transitivity and independence are requirements of rationality "when the preferences are relevant to a sufficiently important decision problem, and where there are no rewards attached to violating transitivity or independence." So in this book I will make it a standing assumption that we are dealing with situations in which the conditions in quotes are satisfied; I can then say simply that transitivity and independence are requirements of rationality.

Suppose, for the moment, that transitivity and independence, and the other conditions necessary for a representation theorem, are indeed requirements of rationality. Then the representation theorem shows that a rational person's preferences maximize expected utility relative to some probability and utility functions. And on the preference interpretation of probability and utility, this vindicates the claim that a rational person has probability and utility functions, and prefers acts that maximize expected utility.

1.4 PREFERENCE

On the understanding of Bayesian decision theory I have been outlining, the notion of preference is central. But what is preference? A behaviorist definition would be: You prefer g to f

12

just in case you are disposed to choose g when presented with a choice between f and g.

Savage (1954, p. 54) rejects a definition like this on the ground that you might be indifferent between f and g, and choose g simply because you must make a choice. However, this is not necessarily a counterexample to the definition; for it may be that you chose g randomly and thus were not *disposed* to choose g. (Analogy: A fair coin does not have a disposition to land heads when tossed, even though it may land heads on some occasions.)

To put the behaviorist definition to a more severe test, suppose you are indifferent between f and g, and are disposed to resolve indifferences by accepting the last option offered. Then if someone offers you f and g in that order, aren't you disposed to choose g over f, without preferring f? If so, this is a counterexample to the behaviorist definition. But in fact the definition can be made to handle this sort of case correctly. We might say that in the case imagined, what you are disposed to choose is *the last option offered*; this option happens to be g, but dispositions may be held to be intensional, in which case it does not follow that you are disposed to choose g.

Nevertheless, there are counterexamples to the behaviorist definition of preference. Suppose you are indifferent between f and g but know that you will shortly face a choice between them. Thinking about the choice in advance, you may decide to choose g, for no reason other than that you must choose something. Having made that decision, you are now disposed to choose g, though you are still indifferent between f and g.[10]

For another sort of counterexample, consider a quiz show in which contestants are asked to select a box, and the boxes differ in nothing other than the number written on them. We might establish (perhaps on the basis of repeated choices) that one contestant is disposed to pick

box 1 when the choice is between boxes 1 and 2;
box 2 when the choice is between boxes 2 and 3;
box 3 when the choice is between boxes 1 and 3.

---

[10]What has been decided on here is a simple plan. The role of plans in constraining future choices is discussed by Bratman (1987).

13

These dispositions are intransitive, and we might criticize the contestant for having intransitive preferences. But the contestant may well respond: "I don't have intransitive preferences. I don't care which box I choose; they are all the same. It just happens that I always make the choices the same way, but surely rationality does not require me to choose sometimes one way, sometimes another, when I am indifferent between options." It seems to me that we should accept this answer; that is, we should accept that the contestant is really indifferent, and thus that the contestant's dispositions to choose do not reflect a preference for one box over the other.

As this last example shows, the difficulty with the behaviorist definition of preference is not merely that it fails to capture the ordinary meaning of 'preference.' There is also the problem that if we adopt the behaviorist definition, then it need not be irrational to have intransitive preferences. Since Bayesian decision theory does deem intransitive preferences irrational, the behaviorist definition is one that Bayesian decision theorists must reject.

Let us use the notation '$f \prec g$' to denote that g is strictly preferred to f, and '$f \sim g$' to denote that the person is indifferent between f and g. These two relations can be defined in terms of weak preference, as follows.

Definition 1.1. $f \prec g \overset{\text{def}}{=} f \precsim g$ and $g \not\precsim f$.

Definition 1.2. $f \sim g \overset{\text{def}}{=} f \precsim g$ and $g \precsim f$.

The point of the preceding paragraphs can then be put by saying that a disposition to choose g is consistent with both $f \prec g$ and $f \sim g$. For this reason, the notion of preference is finer grained than that of disposition to choose.

Although I have rejected the claim that disposition to choose is a *sufficient* condition for preference, I do wish to stipulate that a disposition to choose is a *necessary* condition for preference. That is to say, if $f \prec g$, then you are disposed to choose g when the available options are f and g. Hence if $f \prec g$ at the time of choosing between f and g, you will choose g. This stipulation forces a fairly tight connection between preference and disposition to choose, without actually reducing the one notion to the other.

14

Furthermore, although disposition to choose is not sufficient for preference, it is strong prima facie evidence for preference. And in cases where it is doubtful that a disposition to choose represents a preference, there are ways of obtaining further relevant evidence. For example, we know that most people prefer more money to less. From this, we can infer that if f^+ is like f except that with f^+ the person receives an additional monetary prize, then (ceteris paribus) $f \prec f^+$. Thus if the person is disposed to choose g over f^+, we have that $f \prec f^+ \precsim g$, and assuming that transitivity holds, we can infer that $f \prec g$. On the other hand, if the person chooses g over f but chooses f^+ over g, no matter how small the monetary prize, this would be strong evidence that $f \sim g$.

People sometimes make one choice but say they want to choose differently. A common example is provided by smokers who say they want to quit; in a recent survey, 80 percent of smokers said this.[11] My stipulation about the connection between preference and choice implies that someone who chooses to smoke cannot prefer quitting to smoking. Furthermore, by what was said in the preceding paragraph, the fact that smokers pay money to smoke is evidence that they strictly prefer smoking to quitting. Using obvious notation, we have

$$\bar{s} \prec s. \tag{1.1}$$

How then are we to interpret the smokers who say they want to quit?

One possibility would be to say that for these smokers, quitting is not an option they can choose. If this were so, then the would-be-reformed smokers do not *choose* smoking over quitting, and we could say that for these smokers, $s \prec \bar{s}$. This may be an acceptable account for some of the more deeply addicted smokers, but for the most part we are inclined to say that smokers do have the option of quitting. So we have reason to look for a different solution.

Jeffrey (1974) suggested that a would-be-reformed smoker could be understood as preferring smoking to quitting but also

---

[11] Survey conducted by the Centers for Disease Control; reported in the *New York Times*, July 20, 1989.

15

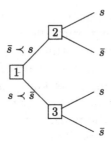

Figure 1.6: Decision between smoking preferences

preferring to prefer quitting. The would-be-reformed smokers' preferences then satisfy both (1.1) and

$$(\bar{s} \prec s) \prec (s \prec \bar{s}). \tag{1.2}$$

We could then say that quitting is an option for these smokers, but preferring quitting to smoking is not an option. Under these circumstances, (1.1) and (1.2) would appear to be consistent and to satisfactorily represent the predicament of the would-be-reformed smokers.

On closer inspection, though, it turns out that (1.2) cannot be attributed to would-be-reformed smokers, on the present conception of preference (which Jeffrey also holds). To see this, suppose Mr. Coffin is given the opportunity to choose between $\bar{s} \prec s$ and $s \prec \bar{s}$. After making that choice, there will be the choice of whether to smoke or not. The choices are represented in the decision tree of Figure 1.6. If Coffin chooses $\bar{s} \prec s$ at node 1, then at node 2 he will choose s. Similarly, if he chooses $s \prec \bar{s}$ at node 1, then at node 3 he will choose $\bar{s}$. Thus his choices at node 1 are equivalent to choosing between s and $\bar{s}$. If he appreciates this equivalence, and if (1.1) holds, then he will choose s in this choice; thus he will choose $\bar{s} \prec s$ at node 1, contrary to (1.2). Thus if we attribute both (1.1) and (1.2) to Coffin, we must also regard him as so confused that he does not appreciate that choosing $s \prec \bar{s}$ is equivalent to choosing $\bar{s}$, while choosing $\bar{s} \prec s$ is equivalent to choosing s. But would-be-reformed smokers are surely not all guilty of a confusion of this sort.

16

My analysis assumes that at node 1, the value Coffin places on s and $\bar{s}$ is the same whether he obtains them via node 2 or node 3. This assumption may be thought objectionable. If Coffin gets $\bar{s}$ via node 3, then he will have reversed his preference to $s \prec \bar{s}$; wouldn't $\bar{s}$ be more desirable in that situation than if $\bar{s} \prec s$? But this worry is misplaced. Preference is a relation between acts; thus s and $\bar{s}$ must be acts. Hence s and $\bar{s}$ must be sufficiently specific that they together with the true state determine what consequence Coffin will obtain; furthermore, the states are assumed to be outside his influence (Section 1.1). Hence s and $\bar{s}$ must be sufficiently specific that Coffin's choice at node 1 does not alter the desirability of s or $\bar{s}$ for him.

I've argued that we cannot represent the would-be-reformed smokers' predicament by ascribing to them (1.1) and (1.2). A natural response is then to try to understand their situation in terms of some other preferences. Here is one attempt: Let $q_{\bar{w}}$ represent the act of quitting without suffering withdrawal symptoms. Of course, this option is not actually available. What is available is the option of quitting and suffering withdrawal symptoms, which we continue to denote q. Then we might try saying that the would-be-reformed smoker's predicament is that $q \prec s \prec q_{\bar{w}}$, but only q and s are available.

However, some would-be-reformed smokers think that they really ought to quit; that is, they think they ought to choose q over s, though they actually choose s instead. This is a case of weakness of will, and I do not think it can be represented in terms of preferences alone. Instead, I suggest the following analysis: When smokers say that they *ought* to quit, they are not expressing a preference but rather are expressing their acceptance of a norm that requires them to quit. Following Gibbard (1990), I take acceptance of a norm to be a mental state that is closely tied to linguistic assertion and also tends to motivate actions; however, the motivational component is not so strong as to preclude the possibility that one will sincerely accept a norm that requires q but nevertheless choose s. So what I would say about these would-be-reformed smokers is that they prefer s to q but accept a norm that requires them to choose q; hence they accept a norm that requires them to have different preferences than those they do have.

Generalizing from this example, I suggest that weakness of will arises when a person chooses in a way that is inconsistent with the norms the person accepts. Thus we can make sense of weakness of will while still insisting that people always choose what they prefer and without having to invoke second-order preferences.[12]

I've now discussed how preference, as I want to conceive of it, is and is not related to choice; but I have not yet said what preference *is*. Work by Davidson, Hurley, and others suggests that for a person to have a preference is for the person to be assigned that preference under the best interpretation of the person. (If there is more than one "best" interpretation, the person has the preference iff[13] it is assigned by all the best interpretations.) If this is accepted, then to say what preference is, I only need to fill in the criteria for evaluating attributions of preference by an interpretation. Much of what I have already said bears on this. Thus I would say that a necessary criterion for a satisfactory interpretation of preference is that it never assign to a person the preference $f \prec g$ at the same time that the person is not disposed to choose g in a choice between f and g. Interpretations satisfying this necessary condition are evaluated by the degree to which they satisfy the following desiderata:

(i) When the person is disposed to choose g over f, the person strictly prefers g to f.

(ii) The person's preferences are normal ones for people to have in the circumstances that the person is in.

(iii) The person's preferences are what the person says they are, except where we have reason to think the person mistaken or insincere.

(iv) The person's preferences are rational.

---

[12]My account of weakness of will agrees with Gibbard (1990, pp. 56–61). Davidson (1980, Essay 2) attempted to give an account of weakness of will that preserves the principle: "If an agent judges that it would be better to do x than to do y, then he wants to do x more than he wants to do y." In my terms, this principle is most naturally read as saying that if you accept a norm that requires choosing x over y, then you prefer choosing x over y. This is precisely what I think we must reject. If I'm right, then Davidson's account of weakness of will must be defective. For an argument that it is, see (Hurley 1989, pp. 131–5).

[13]I follow the common practice of using 'iff' as an abbreviation for 'if and only if'.

In the earlier example of the quiz show contestant, attribution of intransitive preferences violates (ii)–(iv), and my suggestion was that we get a better interpretation by allowing that the person is indifferent between the boxes, thus violating only (i).

1.5 CONNECTEDNESS

Everyone must have had the experience of agonizing over a decision, not knowing what to do. In such a situation, it seems most natural to say that we neither prefer one option to the other nor are indifferent between them. But then we violate Savage's connectedness postulate, since we are faced with options f and g, such that neither $f \precsim g$ nor $g \precsim f$.

Besides the difficult decisions we actually face, there is a huge class of difficult decisions we have not had to face. In many cases, we also lack preferences about the options in these hypothetical decision problems. However, Savage's connectedness postulate requires that one have preferences, even about merely hypothetical options.

Now the axioms of representation theorems are meant to be requirements of rationality, not descriptions of real people. And perhaps it can be argued that when you must choose between some options, rationality requires you to acquire a preference (or indifference) between the options. But it is hard to see how rationality could require you to have preferences also about all the merely hypothetical options that are not available to you.

There is a way to sidestep this worry about connectedness. We could simply *stipulate* that $f \precsim g$ means you do not strictly prefer f to g.[14] Since you cannot strictly prefer both f to g, and g to f, it follows that at least one of $f \precsim g$ and $g \precsim f$ must hold; hence the connectedness postulate is necessarily satisfied. Taking this line means that we cease to distinguish between cases where you are indifferent between f and g, and cases where you lack a view about the relative merits of f and g.

This defense of connectedness only shifts the underlying problem, and does not solve it. To see this, suppose you are undecided between f and g, and let $g + d$ be an act that is just

---

[14]In fact, this is the interpretation that Savage (1954, p. 17f.) gives to the weak preference relation.

19

like g except that you also get a small additional reward d. For example, f and g might be two possible job offers that you would have trouble deciding between, and d might be \$5 (so that $g + d$ is accepting the second job offer and also receiving an additional \$5). If d really is regarded by you as a reward, then $g \prec g + d$. Since you do not strictly prefer g to f, the view we are considering has it that $f \precsim g$. But if your indecision between f and g is at all serious, and if the reward d is sufficiently small, you will also be undecided between f and $g + d$; on the view we are considering, that implies $g + d \precsim f$. So we have $g + d \precsim f \precsim g \prec g + d$; thus you violate transitivity. Yet you need not be irrational to be in this situation, especially if f and g are options you are unlikely to actually have available to you. So if we were to make connectedness an analytic truth, we could no longer say transitivity was a requirement of rationality.

Consequently, I will not interpret $f \precsim g$ as meaning that you do not strictly prefer f to g. Instead, I will continue to interpret it as meaning that you either strictly prefer g to f or are indifferent between them. Since you might be in neither of these states, and since that need not be irrational, I then have to say that the connectedness postulate is not a requirement of rationality.

A more plausible condition is that rationality requires your preferences, so far as they go, to agree with at least one connected preference ordering that satisfies transitivity, independence, and the other assumptions of a representation theorem (Skyrms 1984, ch. 2). I will adopt this condition.

A representation theorem like Savage's shows that if your preferences are connected, and if you satisfy the other assumptions, then your preferences maximize expected utility relative to some probability and utility functions. Furthermore, the theorem shows that the probability function is unique, and the utility function is unique in the sense that if u and v are two utility functions representing your preferences, then there exist constants a and b, $a > 0$, such that $v = au + b$. Now suppose your preferences are not connected but satisfy all the other assumptions of a representation theorem. Then your preferences agree with more than one connected preference ranking that satisfies all the assumptions of a representation theorem. The representation theorem tells us that each of these connected preference rankings is representable by a pair of probability and utility functions. We can then regard your unconnected preferences as represented by the *set* of all pairs of probability and utility functions that represent a connected extension of your preferences. I will call this set your *representor*[15] and its elements *p-u pairs*.

From the way the representor has been defined, it follows that if you prefer f to g, then f has higher expected utility than g, relative to every *p-u* pair in your representor; and if you are indifferent between f and g, then f and g have the same expected utility relative to every *p-u* pair in your representor. You lack a preference between f and g if the *p-u* pairs in your representor are not unanimous about which of these acts has the higher expected utility.

At the beginning of this chapter, I described Bayesian decision theory as holding that rational persons have a probability and utility function, and that a choice is rational just in case it maximizes the chooser's expected utility. In the light of the present section, that description needs a gloss. The statement that a rational person has a probability and utility function is not meant to imply that these functions are unique; a more explicit statement of the position I am defending would be that a rational person has a representor that is nonempty; that is, it contains at least one *p-u* pair. A corresponding gloss is needed for the statement that a choice is rational just in case it maximizes expected utility; a more complete statement of this condition would be that a choice is rational just in case it maximizes expected utility relative to every *p-u* pair in the chooser's representor.

From this perspective we can say that although connectedness is not a requirement of rationality, and although representation theorems assume connectedness, nevertheless representation theorems do provide the foundations for a normatively correct decision theory.

1.6 NORMALITY

The preference relations $f \precsim g$, $f \prec g$, and $f \sim g$ only deal with your attitude to choices in which there are just the two options

---

[15] The term comes from van Fraassen (1985, p. 249), but the meaning here is different.

21

f and g; they imply nothing about your attitude when there are more than these two options. To see this, note that if $f \prec g$, then according to the stipulations I have made, in a choice between f and g you would choose g; however, it is consistent with this that in a choice where the options are f, g, and h, you would choose f. In order to be able to talk about your attitude to decision problems with more than two options, I will introduce a generalization of the binary preference relations $f \precsim g$, $f \prec g$, and $f \sim g$.

Let '$\mathcal{C}(F)$' denote the acts you prefer or want to choose when the set of available acts is F. For example, if you prefer to choose f when the available acts are f, g, and h, then $\mathcal{C}\{f,g,h\} = \{f\}$. If you prefer not choosing h but are indifferent between choosing f and g, then $\mathcal{C}\{f,g,h\} = \{f,g\}$. If you are indifferent between all three options, then $\mathcal{C}\{f,g,h\} = \{f,g,h\}$. If you are undecided about what to choose, then $\mathcal{C}\{f,g,h\}$ is the empty set, denoted by $\emptyset$. The function $\mathcal{C}$ so defined is called your *choice function*, and $\mathcal{C}(F)$ is called your *choice set* for the decision problem in which F is the set of available options. Of course, your choice function will in general change over time.

The notion of a choice function generalizes the binary preference relations. To see this, note that the binary preference relations can be defined in terms of the choice function, as follows:

$f \precsim g$ iff $g \in \mathcal{C}\{f,g\}$;
$f \prec g$ iff $\mathcal{C}\{f,g\} = \{g\}$;
$f \sim g$ iff $\mathcal{C}\{f,g\} = \{f,g\}$.

Thus in an axiomatic presentation of this subject, one would begin with the choice function and then define the binary preference relations. I have followed the reverse order because I think it is easier to first master the simpler concepts of binary preference and then generalize to the choice function.

The discussion of the meaning of preference, in Section 1.4, generalizes in the obvious way to choice functions. In particular, the choice function is to be understood as satisfying the following stipulation:

If $\mathcal{C}(F) \neq \emptyset$ for you at the time of deciding between the acts in F, then you will decide on an act in $\mathcal{C}(F)$.

This implies, in particular, that if $C(F) = \{f\}$, you will choose f when the set of available acts is F.

Part of what is asserted by Bayesian decision theory cannot be expressed in terms of binary preferences, but rather requires the notion of a choice function. To see this, suppose your binary preference relation $\precsim$ satisfies all the requirements of Bayesian decision theory (transitivity, independence, etc.); and suppose that for you $f \prec g$, but $C\{f, g, h\} = \{f\}$. Since $f \prec g$ and your preferences accord with Bayesian decision theory, g must have higher expected utility than f, in which case f does not maximize expected utility in $\{f, g, h\}$, and so according to Bayesian decision theory, choosing f is irrational in this decision problem. Thus your choice function is inconsistent with Bayesian decision theory, even though your binary preferences are consistent with it.

Since the assumptions made in representation theorems concern binary preference only, we see that Bayesian decision theory is committed to *more* than the assumptions that appear in representation theorems. This extra commitment can be stated as follows.

Normality. *If F is a set of acts on which $\precsim$ is connected, then*

$$C(F) = \{f \in F : g \precsim f \text{ for all } g \in F\}.$$

(The term 'normality' comes from [Sen 1971].) It is easy to verify that this condition must hold if Bayesian decision theory holds. For according to Bayesian decision theory, $f \in C(F)$ iff f maximizes expected utility in F; that is, iff $EU(g) \leq EU(f)$ for all $g \in F$; and that is equivalent to $g \precsim f$ for all $g \in F$.

In Chapter 2, I will defend the view that normality is a requirement of rationality.

1.7 RATIONALITY

Bayesian decision theory, as I have been articulating it, is concerned with the rationality of preferences and choices. But what is rationality?

In many discussions of Bayesian decision theory, the theory appears to be taken as holding that maximization of expected utility is (at least part of) what it *means* to be rational. If this

23

is accepted, then the claim that rational persons maximize expected utility becomes a tautology. But it seems clear that this is not a tautology, as 'rationality' is normally understood. That objection could be avoided by taking the theory to be giving a stipulative definition of 'rationality' rather than attempting to capture the ordinary meaning of the term; but then we would need some argument why rationality in this sense is an interesting notion. Such an argument would presumably try to show that we *ought* to maximize expected utility – in other words, that maximizing expected utility is *rational*, in a sense other than the one just stipulated. My topic in this section is what that other, essentially normative, sense of 'rationality' might be.

Taylor (1982) suggests that irrationality is just inconsistency (though he denies that rationality is just consistency). Davidson (1985) works with a definition of irrationality as violation of one's own standards. And Foley (1987) favors the view that a person's belief is rational if it conforms to the person's deepest epistemic standards. However, I do not think that this (in)consistency conception of (ir)rationality serves all the legitimate purposes for which we use the notion of (ir)rationality. An example that is particularly relevant to this book: Philosophers of science are concerned with normative theories of scientific method. One way to evaluate such theories is to test them against our judgment of particular cases. Thus we might, for example, have a firm pretheoretic intuition that geologists in the 1960s were rational to accept continental drift. If a theory of scientific method agrees with this intuition, that is a point in favor of the theory; otherwise, we have a strike against the theory. But when we judge what was rational for geologists in the 1960s, we are not primarily concerned with whether accepting continental drift was consistent with their standards, however interesting this question may be. For the purposes of evaluating our theory of scientific method, what we need to judge is whether we think geologists in the 1960s *ought* to have accepted continental drift. The 'ought' here involves a notion of rationality that goes beyond consistency.

The notion of rationality we need here is, I think, best analyzed along the lines of the norm-expressivist theory in Allan Gibbard's recent book (1990). On this theory, we do not

24

attempt to fill in the blank in "'X is rational' means ...."
Rather, we say what someone is doing when they call X rational. And the theory says that when you call something rational, you are expressing your acceptance of norms that permit X. So, for example, to say geologists in the 1960s were rational to accept continental drift is to express your endorsement of norms that permit acceptance of continental drift in the circumstances geologists were in in the 1960s.

On Gibbard's view, to say that something is rational is not to make a statement that could be true or false; in particular, the assertion about geologists in the 1960s does not say that their acceptance of continental drift was consistent with the norms of either the speaker or the geologists. However, this assertion does *express* the speaker's acceptance of norms that permit these geologists to accept continental drift.

What I have said is just the barest outline of an analysis of rationality. For completeness, it especially requires an account of what it is to express one's acceptance of a norm. Here I refer the reader to Gibbard's discussion, which admirably presents one way such an account may be developed.

1.8 JUSTIFICATION

We talk not only about the *rationality* of actions and beliefs, but also about whether or not they are *justified*. What is the relation between these notions of rationality and justification? A common view is that they coincide; that is, that something is rational just in case it is justified. I shall argue that this is not so, and that an action or belief can be rational without being justified.

Consider the stock example of a man who has good evidence that his wife is unfaithful. If he were to believe her unfaithful, this would show – and probably would bring the marriage to an end; so, all things considered, his interests are best served by believing that she is faithful, contrary to his evidence.[16] It seems clear that this man would not be *justified* in believing his wife faithful. On the other hand, the belief would be in the

---

[16]For definiteness, let 'belief' here mean high probability.

25

man's best interest and would do no harm to anyone else, so it seems that the norms we accept should permit him to believe this, in which case, the norm expressivist theory would have us saying that it is *rational* for the man to believe his wife unfaithful. Thus we seem to be forced to divorce the notions of rationality and justification.

Gibbard has a way of attempting to avoid this divorce. He suggests that it would be rational (and justified) for the man to *want* to believe his wife faithful, but irrational (and unjustified) to actually *believe* it (Gibbard 1990, p. 36f.). Gibbard would allow that what is said here for wanting also applies to preference in my sense. For example, suppose the man can get himself to believe his wife faithful by some means, and let b be the act of doing this. Then Gibbard would allow that the man is rational (and justified) to have $\bar{b} \prec b$, while still maintaining that the man would be irrational (and unjustified) to have the belief that b produces.

Is this a consistent position? In trying to answer that question, I will assume that consistency requires that the norms one accepts should be jointly satisfiable in at least one possible world. This is in accord with Gibbard's normative logic (1990, ch. 5), though I will gloss over certain refinements in Gibbard's account.

I have made it a conceptual truth that someone who has $\bar{b} \prec b$ will choose b when the alternatives are b and $\bar{b}$. So if we accept that the man would be rational to have $\bar{b} \prec b$, we accept that the following norm applies to this man.

(1) If you have a choice between b and $\bar{b}$, choose b.

But b is an act that the man knows will produce in him the belief that his wife is faithful. Given this knowledge, there is no possible world in which the man chooses b but does not believe his wife is faithful. So if we judge that the man would be irrational to believe his wife faithful, we must accept that this norm applies to him:

(2) Do not choose b.

The only worlds in which (1) and (2) are both satisfied are those in which the man does not have a choice between b and

26

$\bar{b}$. Hence Gibbard must conclude that the man should not have the choice between b and $\bar{b}$. That is, Gibbard must accept the following norm:

(3) Do not have a choice between b and $\bar{b}$.

The thought here would be that fully rational persons are not able to get themselves to believe things that are not supported by the evidence. This is a possible position to take, so there is not yet any inconsistency.

Now suppose that our man does have the power to get himself to believe his wife is faithful. For the sake of argument, I will concede that his having this power is a failure of rationality. Still, he has it, and we can suppose that there is nothing he can do about it. A useful system of norms will say what should be done in situations where some norm is unavoidably violated. In a different context, Gibbard discusses norms of this sort, calling them "norms of the second-best" (1990, p. 241). So now I ask: Given that the man is unable to satisfy both (1) and (2), which one should we regard as applying to his situation?

Suppose we decide to endorse (1), and drop (2) as a norm of the second-best. Then we are saying, according to Gibbard, that in this situation the man is both rational and justified to believe that his wife is faithful. But in view of the contrary evidence, the claim that the man is justified in believing this seems plainly false.

So let us instead try endorsing (2), and dropping (1) as a norm of the second-best. Then we are saying, according to Gibbard, that the man would be irrational and unjustified to believe his wife faithful. But there is a clear sense in which it makes sense for the man to believe his wife faithful. Someone who advised the man to acquire this belief would have given the man good advice. In the original situation, Gibbard proposed to accommodate this by saying that wanting to believe is rational; but we are now in a situation where preference entails choice, so that option is not available here, unless we go back to the unsatisfactory situation of the preceding paragraph.

I conclude that Gibbard's attempt to equate rationality and justification, elegant though it is, ultimately fails. I think we have to accept that rationality and justification are different notions. I would say that the man is *rational* to get himself to believe (and hence to believe) that his wife is faithful, but is *not justified* in believing this.

The norm-expressivist analysis is, I think, correct for rationality. That is, I agree that calling something rational is expressing acceptance of norms that permit it. Thus in saying that the man would be rational to believe his wife faithful, I am expressing my acceptance of norms that permit the man to believe this. Because the man is nevertheless not justified in believing this, I think the norm-expressivist analysis does not give a correct account of justification.

Of course, justification is itself an evaluative notion, like rationality; so if a norm-expressivist analysis of rationality is right, one should look for an analysis of justification along the same lines. In the case of subjective probability, we might try the following. The standard way in which subjective probabilities have value or disvalue is via their connection with preference for actions, and hence with action itself. It is desirable to have subjective probabilities that result in successful actions being chosen and unsuccessful ones avoided. Gibbard calls probabilities that are desirable in this sense *systematically apt* (1990, p. 221). But in some cases, subjective probabilities can have value or disvalue in another way. This is what happens in the case of the man with evidence that his wife is unfaithful. The reason he does not want to apportion probability in accordance with the evidence is that this would produce involuntary signs that would destroy the marriage; it is not that the preferences associated with this probability would be inappropriate. Indeed, I will show in Chapter 5 that considering only the influence on actions chosen, the man would maximize expected utility by assigning probabilities in accordance with the evidence. So what we might try saying is that for subjective probabilities to be *justified* is for them to be systematically apt – that is, rational if we consider only the influence on actions chosen and ignore other consequences. Then to call a subjective probability judgment justified is to express one's endorsement of norms that permit it *in circumstances where no extrasystematic consequences attach to having that probability.* A judgment of

28

justification would then be a conditional judgment of rationality, dealing with a possibly counterfactual situation. It is clear why such judgments would be useful.

Even if this account of justification is right as far as it goes, more work would be needed to deal with justification of other things, such as emotions and moral judgments. I will not pursue that here, because this book is about rationality, not justification. For my purposes, it suffices to have established that rationality is not the same thing as justification.

1.9 QUALIFICATION

I began this chapter by saying that according to Bayesian decision theory, it is rational to choose act a just in case a maximizes expected utility, relative to your probabilities and utilities. Taken literally, this position would imply that questions of the rationality of your probabilities and utilities are irrelevant to the question of what you rationally ought to do. It seems that currently most Bayesians accept this literal interpretation of the theory. For example, Eells (1982, p. 5) allows that an action that maximizes your expected utility may not be well-informed, but he does not allow that it could be irrational. I will call this position *unqualified Bayesianism*.

I do not accept unqualified Bayesianism. I think we should allow that probabilities and utilities can themselves be irrational; and when they are, it will not generally be rational to maximize expected utility relative to them. Thus I view the principle of maximizing expected utility as elliptical; to get a literally correct principle, we must add to the principle of expected utility the proviso that the probabilities and utilities are themselves rational. I will call the position I am defending *qualified Bayesianism*.

In saying that probabilities and utilities can themselves be irrational, I am *not* saying that rationality determines unique probability and utility functions that everyone with such-and-such evidence ought to have. On the contrary, I would allow that a wide range of probability and utility functions may be rational for persons who have had all the same experiences. Thus I reject positions like that of Carnap (1950) or Salmon (1967) on what it takes for probabilities to be rational. Qualified Bayesianism is not a species of "objective Bayesianism," as that term has usually been understood.

Then why do I advocate qualified Bayesianism? Let me begin by noting that there is no positive argument for unqualified Bayesianism. In particular, representation theorems do not provide such an argument. What representation theorems show (if their assumptions are accepted) is that you are not fully rational if your preferences do not all maximize expected utility relative to your probability and utility functions. It follows that maximization of your expected utility is a *necessary* condition of rationality. But it does not follow that maximization of your expected utility is a *sufficient* condition of rationality. Nor is there any other positive argument for unqualified Bayesianism, so far as I know. By contrast, representation theorems do provide an argument for qualified Bayesianism. For if it is rational to have p and u as your probability and utility functions, and if (as the representation theorems show) a rational person maximizes expected utility, then it is rational for you to maximize expected utility relative to p and u.

Though there is no positive argument for unqualified Bayesianism, there are arguments against it. One is that even if a person's preferences all maximize expected utility relative to p and u, having these preferences may conflict with the norms that the person accepts, and thus be irrational by the person's own lights. In such a situation, we have to say either that the person's preferences are irrational or the person's norms are mistaken. Unqualified Bayesianism entails that the first option is impossible (since the preferences all maximize expected utility relative to the person's probability and utility functions); thus it entails that the person's norms are mistaken. But in some cases, the opposite conclusion will be at least as plausible.

For example, suppose that I am one of those smokers who accept that they ought to quit; but being weak willed, I go on smoking. According to the account of preference in Section 1.4, I prefer smoking to quitting. Furthermore, my preferences might satisfy transitivity, independence, and the other assumptions of a representation theorem. If so, I have probability and utility functions, and the options I prefer maximize

30

expected utility relative to my probability and utility functions (Section 1.3). Hence smoking maximizes my expected utility. Unqualified Bayesianism concludes that I am rational to go on smoking, and hence that I am in error in thinking that I ought to quit. But it seems at least as plausible to say that my norms are right and it is my preferences that ought to change.

For another sort of example, suppose I prefer to go back into the house rather than cross a black cat in the street. Suppose further that this preference maximizes my expected utility, because I have a high probability for the claim that I will have bad luck if I cross a black cat. Elster (1983, p. 15) says Bayesian decision theory is an incomplete theory of rationality because it does not deem this preference irrational. Unqualified Bayesianism has to say that the preference is not irrational; and I would agree that the preference is not *necessarily* irrational. But suppose that I myself accept that my view about the influence of black cats is irrational, but I have not been able to rid myself of it. Then I think I am irrational, and it seems bizarre to say I must be wrong about this.

A possible response here would be to say that when a person's norms and choices diverge, we should take the norms rather than the choices as determining the person's preferences. But we want to attribute preferences to people on matters about which they have no normative commitments; and we attribute preferences to beasts, who accept no norms at all. So this proposal would make preference a gerrymandered concept. Furthermore, the proposal would not remove the basic problem; for when a person's norms and choices diverge, it is not necessarily the norms that are correct.

Second argument against unqualified Bayesianism: We attribute probabilities and utilities to persons, even when we know that their preferences do not all maximize expected utility relative to any one p-u pair. My account of personal probability in Section 1.3 allowed for this, by saying that p and u are your probability and utility functions if they provide an interpretation of your preferences that is sufficiently good and is better than any competing interpretation. So now, suppose that most of my preferences maximize expected utility relative to p and u, and on this account p and u are my probability and utility functions. But I have some preferences that do not maximize expected utility relative to p and u. Unqualified Bayesianism here endorses majority rule: It says that most of my preferences are rational and the few that diverge are irrational. But when I reflect on the conflict, I might well decide that it is the minority that are right and the majority that are wrong. There seems no reason to think it would always be the other way around.

For example, I might be in the grip of the gambler's fallacy and think that when a coin has been tossed and landed heads, it is more likely that it will land tails the next time. Here a probability is attributed to me because it fits my preferences regarding bets on coin tosses. However, I may have a few other preferences that do not fit. For example, suppose I would bet that there is no causal influence between different tosses of the coin, though such a bet does not maximize expected utility relative to the probability and utility functions attributed to me on the basis of most of my preferences. Unqualified Bayesianism implies that the latter preference is irrational; but in this case I think most of us would say this preference is rational, and it is the ones that maximize my expected utility that are irrational.

Third argument: Suppose that, for no reason, I suddenly come to give a high probability to the proposition that my office is bugged. Most Bayesians will agree that if this new opinion was motivated by no evidence, then I was irrational to adopt it. In the jargon to be introduced in Chapter 4, the shift violates "conditionalization." In Chapter 5, I will show that Bayesians are right about this, provided certain conditions hold. So, let us grant that I was irrational to shift from my former view to the new one that my office is bugged. Question: Now that I have made the shift, is it rational for me to accept an even-money bet that my office is bugged? Unqualified Bayesianism says it is, because accepting the bet maximizes my current expected utility. But even unqualified Bayesians agree that I was irrational to acquire my current probability function; thus, unqualified Bayesianism seems committed to the view that once errors are made, they should not be corrected. It is hard to see the merit in this view; indeed, I suggest that the view is patently false.

Finally, let me draw on an analogy with the principle of deductive closure. This principle is apt to be stated in an unqualified form, according to which it holds that persons ought to accept the logical consequences of what they accept. But on reflection, we soon see that this principle is untenable. For we want to allow that when the consequences of what we accept turn out to be sufficiently implausible, we should not accept them but instead revise some of the things we now accept. Thus the principle is acceptable only if qualified; we should say that *if it is rational to accept what you have accepted*, then it is rational to accept the logical consequences of what you have accepted. This qualified version of the principle of deductive closure is analogous to qualified Bayesianism. The unqualified version of deductive closure, which we quickly saw to be untenable, is analogous to unqualified Bayesianism.

Of course, qualified principles are not as informative as unqualified ones. Nevertheless, they can assist in making decisions. Consider first the qualified version of deductive closure. If you find that a proposition is a consequence of those you have accepted, the principle tells you that either the proposition should be accepted or something you have accepted should be abandoned; and though it does not tell you which to do, you will commonly be able to make a judgment in favor of one option or the other. The contribution that qualified Bayesianism can make to resolving decision problems is similar. You can determine which options in a decision problem maximize your expected utility, and qualified Bayesianism then tells you that you should either choose one of those options or else have different probabilities or utilities. Normally you will be able to make a judgment in favor of one or other of these alternatives. Here again, as in Section 1.2, we see that Bayesian decision theory is an aid to rational decision making though not an algorithm that can replace good sense.

I turn now to the argument that transitivity, normality, and independence are requirements of rationality. The reader who needs no convincing on this matter could skip the next two chapters.

33

2

Transitivity and normality

In this chapter, I begin by noting that transitivity and normality enjoy wide endorsement. I then consider arguments for transitivity and normality, and give reasons for thinking that these do not provide additional reason to endorse transitivity and normality. Then I consider arguments against transitivity and normality, and give reasons for thinking that they too are ineffective. I conclude that transitivity and normality are about as secure as any useful principle of rationality is likely to be.

2.1 POPULAR ENDORSEMENT

Although violations of transitivity and normality are not uncommon, most people, on discovering that they have made such a violation, feel that they have made a mistake. This indicates that most people already regard transitivity and normality as requirements of rationality.

Experimental evidence for this was obtained by MacCrimmon (1968). MacCrimmon's subjects were business executives in the Executive Development Program at the University of California at Los Angeles. In one of these experiments, subjects were asked to assume that they were deciding on the price of a product to be produced by their company. They were also asked to assume that four of the possible pricing policies would have the following consequences.

A. Expected return   10%
    Expected share of market   40%

B. Expected return   20%
    Expected share of market   20%

C. Expected return   5%
    Expected share of market   50%

34

D. Expected return   15%
    Expected share of market   30%

The subjects were presented with all possible pairs of these options and asked to indicate a preference. Options from two other transitivity experiments were interspersed with the presentations of these options.

In this experiment, only 2 of 38 subjects had intransitive preferences. Altogether, in the three experiments (which all had a similar structure), 8 of the 114 subject-experiments exhibited intransitivity. After the experiment, subjects were interviewed and given a chance to reflect on their choices. MacCrimmon writes:

During the interview, 6 of these 8 subjects quickly acknowledged that they had made a "mistake" in their choice and expressed a desire to change their choices. The remaining 2 subjects persisted in the intransitivity, asserting that they saw no need to change their choices because they focused on different dimensions for different choice pairs. The fallacy of this reasoning was not apparent to them in a five-minute discussion.

Thus MacCrimmon's results support the view that an overwhelming majority of people regard transitivity as a requirement of rationality; but it also shows that some people do not.[1]

After making the pairwise comparisons, subjects in MacCrimmon's experiment were asked to rank order all four options. MacCrimmon refers to a discrepancy between the rank ordering and the pairwise comparisons as a "choice instability." Any violation of normality would be a choice instability, but the converse does not hold. MacCrimmon reports:

Only 8 of the 38 subjects had no "choice instabilities" in the three experiments. The other 30 subjects (except for the 2 who persisted in intransitivities) all wished to change their choice (most would change the binary choice, but the difference was not significant). They generally attributed the "choice instability" to carelessness in their reading or thinking.

[1] Tversky (1969) obtained a higher rate of intransitivity than occurred in MacCrimmon's study, but he too found that "the vast majority" of subjects regarded transitivity as a requirement of rationality.

35

MacCrimmon's results thus provide evidence that commitment to normality is at least as widespread as commitment to transitivity. There are, however, the holdouts. Furthermore, some of those who reject transitivity and normality are well-informed students of decision theory. For that reason, arguments have been offered that attempt to derive transitivity, and sometimes normality also, from more compelling principles. I discuss such arguments in the next section.

2.2 ARGUMENTS FOR TRANSITIVITY AND NORMALITY

I first discuss the best-known argument for transitivity, the money pump argument. I show that this argument is fallacious. I then consider three other arguments, which I take to be sound, but whose premises seem to me no more plausible than transitivity and normality themselves. Thus my ultimate conclusion is that none of these arguments adds much to the case for transitivity and normality. However, this is not a ground for doubting transitivity and normality; it rather reflects the difficulty of finding premises more plausible than transitivity and normality.

2.2.1 The money pump

The money pump argument is presented by Davidson, McKinsey, and Suppes (1955), Raiffa (1968), and in many other places. It goes like this:

Suppose you violate transitivity; then there are acts f, g, and h such that $f \precsim g \precsim h \prec f$. If you had h, then since $h \prec f$, you should be willing to pay a small premium to exchange h for f. But then, since $f \precsim g \precsim h$, you should be willing to exchange f for g, and then g for h. But now you are back where you were, except poorer by the premium you paid to exchange h for f. Furthermore, we could go around the cycle repeatedly, collecting a premium from you each time you exchange h for f. Thus you are said to be a *money pump*.

This is such a simple and vivid argument that it is a pity it is fallacious. But fallacious it is. The fallacy lies in a careless analysis of sequential choice. The argument assumes that someone with intransitive preferences will make each choice without any thought about what future options will be available, yet this is

36

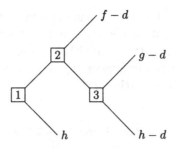

Figure 2.1: The money pump decision tree

not in general a rational way to proceed. For example, suppose $f \precsim g \precsim h \prec f$, that you now have h, and that you have the opportunity to exchange h for f, at a small cost d. Suppose you know that if you make the exchange, you will then be offered g for f; and subsequently h for g. Then your decision problem is really the sequential decision problem represented in Figure 2.1. In this tree, the nodes (which are numbered) represent points at which you have to make a decision. Node 1 represents the point at which you decide whether to exchange h for f, and pay d for the privilege. If you do not make the exchange, you get h; otherwise you move to node 2. At node 2 you have to decide whether to exchange f for g; if you do not make this exchange, you get $f - d$; otherwise, you move to node 3. At node 3, you have to decide whether to exchange g for h.

The money pump argument assumes that since $g \precsim h$, then $g - d \precsim h - d$, and hence that you would be willing to exchange g for h at node 3. But if you knew you would do this, then the choice at node 2 is effectively between $f - d$ and $h - d$. And the money pump argument assumes that since $h \prec f$, it is also the case that $h - d \prec f - d$. Consequently, at node 2 you ought to choose $f - d$, and hence refuse to exchange f for g. The cycle, envisaged by the money pump argument, is broken at this point.

This analysis assumes you know what choices will be offered to you in the future. If this were not known, then you might well make choices that would leave you worse off than when you started. But this fact does not provide much of an argument against intransitivity. An investor who buys stocks and

37

later sells them at a lower price is not necessarily irrational; things have just not turned out the way they were expected to. Similarly, persons with intransitive preferences may argue, the fact that they might unexpectedly end up with a loss does not show that they are irrational.[2]

2.2.2 Consequentialism

Consequentialism is the principle that the value of acts is determined by their consequences in the various possible states of the world. Hammond (1988a) argues that this principle suffices to derive many of the postulates of a representation theorem, including transitivity and normality. Hammond's argument is like the money pump argument in that it considers sequential choices over time, but Hammond does not make the assumption of myopia that vitiated the money pump argument. As presented by Hammond, the argument is couched in formidable notation; I will try to state it in English.

Hammond begins by laying down a principle that I will call *rigidity*. (Hammond calls it "consistency.") This principle requires that the choices a person would prefer on reaching any choice node in any decision tree are the same as those the person would now prefer if now faced with the choices available at that node. Hammond claims that rigidity is a very weak principle, and to support that claim he discusses the following example: A potential drug addict faces the decision of whether or not to abstain entirely from a drug; and if the drug is tried, then there is the decision of whether or not to quit. Figure 2.2 represents the situation. We can suppose that initially (at node 1), what the potential addict most prefers is to use the drug, then quit before permanent damage is done. However, by the time node 2 is reached, the potential addict will have become an actual addict and will choose to continue use. Still, Hammond argues, rigidity is satisfied in this example. For he claims (p. 36, line 1)

---

[2]Lehrer and Wagner (1985) argue for transitivity by showing that in a particular example, if you have intransitive preferences, then the following holds: If you make sequential choices myopically in accordance with those preferences, then you may end up with an option that you disprefer to some other option that was available. To infer from this that intransitivity is irrational is to make the same sort of error as I have just identified in the money pump argument.

38

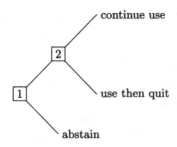

continue use

2

use then quit

1

abstain

Figure 2.2: The potential addict's problem

that the potential addict would now choose to continue use if now faced with the choices at node 2.

The rationale for the latter claim is presumably that someone faced with the choices at node 2 has already tried the drug, and hence the potential addict faced with these choices must be addicted. But this need not be so. Suppose the potential addict has never tried the drug, is offered the choice between a small and a large quantity of it, has no other possibility of procuring the drug, and is required to consume at least the small quantity. Then the choices are those at node 2, but the potential addict is not yet addicted; and so we can expect that the potential addict would choose the small quantity (which amounts to choosing to use and then quit). But still, we can suppose, it remains the case that at node 2 in the tree in Figure 2.2, the potential addict will have become an actual addict and thus will prefer continued use. Then contrary to what Hammond claims, rigidity is violated in this example.

So rigidity is a substantive assumption, and the question becomes whether it is plausible as a principle of rationality. The fact that we are all potential addicts, and are not irrational on that account, suggests that rigidity is not a requirement of rationality. Perhaps it will be urged that an actual addict has irrational preferences, and this is the source of the failure of rigidity. But if this defense is relied on, then Hammond's argument would only show that transitivity and normality are satisfied by those who not only are rational, but also would be rational under all future contingencies. Since none of us are in that position, this

39

would leave open the possibility that transitivity and normality are not requirements of rationality for us.

In any case, potential addiction is not the only way in which normal people violate rigidity. Consider a ten-year-old boy who prefers that he never get involved with women, but who will have very different preferences in a few years' time. Here there is a failure of rigidity, but I don't think we want to say that either the ten-year-old or the sixteen-year-old is irrational. A possible response to this would be to say that the trees in Hammond's argument can be limited to ones that will be completed in a short period of time, so that significant maturation cannot occur within the tree. But do we really want to make the defense of transitivity and normality rest on contingent assumptions about the rate of human maturation?

There are other objections that can be raised against rigidity, and I will present some of them in Section 3.2.2. I don't press them here, because I guess that Hammond would respond to any one in a way that meets them all. Hammond remarks that "the potential addict is really two (potential) persons, before and after addiction" (p. 36). If we are using usual criteria of personal identity, then this statement seems clearly false; addiction does not in general destroy personal identity. But the statement can alternatively be interpreted as stipulating an idiosyncratic notion of personal identity, according to which drug-induced changes in preferences destroy personal identity. If we are going to take this step, we might as well go the whole way and stipulate that any change of preference over time destroys personal identity. Thus all objections to rigidity are met. Rigidity then becomes, not merely a very weak principle (as Hammond claims), but an empty tautology.

In addition to rigidity, Hammond adopts another principle, which he says is implied by consequentialism. The principle is this: The consequences you can obtain in a decision tree, choosing at each node in accordance with your preferences at that node, depend only on the consequences available in the tree and thus are independent of the structure of the branches in the tree. I will call the latter principle *path independence*. Hammond shows that rigidity and path independence together entail transitivity and normality.

40

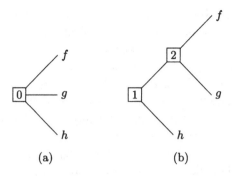

Figure 2.3: Trees to illustrate (∗)

If rigidity is assumed to hold, then the person referred to in the path-independence principle is one whose preferences do not change as the person traverses the tree. What path independence then amounts to is this:

(∗) The options you can obtain in a decision tree, by choosing at each node in accordance with the preferences you now have for the options at that node, are a function of the consequences available in the tree.

I think Hammond is most illuminatingly read as showing that (∗) entails transitivity and normality.

To see why this entailment holds in a particular case, suppose you violate normality by having $f \prec g$, but $C\{f, g, h\} = \{f\}$. Then in Figure 2.3 tree (a), choosing in accordance with your preferences would give you f. But in tree (b), choosing in accordance with your current preferences at node 2 would give you g; thus you cannot obtain f by choosing in accordance with the preferences you now have at each node in tree (b). Since the same consequences are available in trees (a) and (b), this is a violation of (∗).

Is this violation of (∗) patently irrational? Someone who rejects normality might claim that there is nothing irrational about having the options one can obtain depend on the structure of the decision problem. Alternatively, the opponent of normality could concede that the options a person can choose should not depend on the structure of the problem, but could

41

deny that this shows violations of normality are irrational; instead, it could be maintained that what the case shows is that rational persons should sometimes change their preferences as they move through a decision tree. For example, in tree (b) of Figure 2.3, this view would hold that what you should do is proceed to node 2, and as you do so change your preferences to have $g \prec f$, thus ensuring that you choose f. McClennen (1990) would make this second response.

Now we may well claim that these objections to $(*)$ are mistaken. But that claim seems to me no more obvious than the claim that transitivity and normality are requirements of rationality. So while I personally would endorse $(*)$, I do not think we add to the credibility of transitivity and normality by deriving them from $(*)$. In other words, I judge Hammond's argument to be sound but not useful.

2.2.3 Modesty

I now present another argument based on consideration of sequential choice situations. The premise of this argument may seem more compelling than Hammond's, at least initially. Though the argument can be generalized to derive both transitivity and normality, I here keep things simple by considering it only as an argument against the most extreme type of violation of transitivity.

Let us say that your preferences are *strictly intransitive* if for some f, g, and h, you have $f \prec g \prec h \prec f$. If your preferences are like this, then in certain sequential decision problems, you yourself would prefer to have different preferences to those you actually have. For example, consider the decision problem represented in Figure 2.4. Since $f \prec g$, you may anticipate that you would choose g at node 2, in which case your choice at node 1 is effectively between g and h. Since $g \prec h$, you would then choose h. But now consider what would happen if you had $f \succ g$ while other preferences remain unchanged. Then you anticipate that at node 2 you would choose f; hence your choice at node 1 is effectively between f and h; and so, since $h \prec f$, you would choose to proceed to node 2 and obtain f. Thus with your present preferences, you would obtain h; but by reversing the preference between f and g you would obtain f. Since

42

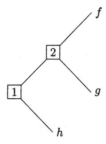

Figure 2.4: Tree to show strict intransitivity entails modesty

$f \succ h$, you therefore prefer that in this decision problem your preference be different from what it is. We can describe this by saying that your preferences are *modest*, since they themselves deem different preferences better. And this seems to be a failure of rationality.[3]

Note, though, that this argument assumes that you know your preference between f and g will be the same when you reach node 2 as it is now. If you knew you would at node 2 reverse your preference between f and g, you would not prefer to now have different preferences, and thus your preferences would not be modest. Thus an opponent of transitivity could agree that someone with modest preferences is irrational but claim that in the present example the irrationality derives, not from the violation of transitivity, but from the failure to change preferences appropriately when moving from node 1 to node 2. This would be the position of McClennen (1990).

Modesty is the property that you prefer that your *current* preferences be different. What I have just noted is that strict intransitivity entails modesty only if you know that your preferences

---

[3]The following might appear to be a counterexample to this claim: If I preferred buying XYZ Corporation's stock tomorrow to not buying, I would have evidence that the stock will rise, and thus I would be likely to make a profit. But in fact, I don't have such evidence and prefer not buying the stock. Thus I seem to prefer having different preferences to those I actually have, without being irrational. But here what I prefer is that I change my preferences as a result of a learning experience; if it were within my power to just choose to have different preferences, then absent any learning, I prefer to leave my preferences as they are. By contrast, the case described in the text is one where you would want to choose to have different preferences, without any learning. I mean to refer only to situations of the latter kind when I call preferences modest.

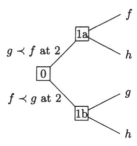

Figure 2.5: Choice of future preferences

at future nodes will be the same as your current preferences for the options available at those nodes. Thus modesty by itself cannot be used to argue for transitivity. However, we might try this: Say that your preferences are *diachronically modest* if you now prefer that at some future node in a decision problem, your preferences would be different from your current preferences. It might seem that diachronic modesty is also a failure of rationality, and that the argument so far shows that strict intransitivity entails diachronic modesty.

The main problem with this argument is with the notion of preferences for future preferences. To interpret such preferences, we need to connect them with choice. Presumably, to say "you now (prior to choosing at node 1) prefer that at node 2 you prefer f" implies this: If you could now choose what preference you would have at node 2, and then make your choice at node 1, you would choose to prefer f. But then the decision tree we are considering becomes the one shown in Figure 2.5. Here the choice of what preference to have at node 2 is made at node 0, and nodes 1a and 1b correspond to node 1 in Figure 2.4. If at node 0 you anticipate choosing in accordance with your current preferences at node 1 (a or b), then you anticipate that choosing to proceed to node 1a will result in obtaining f, while choosing to proceed to node 1b will result in obtaining h. Since $h \prec f$, you will therefore proceed to node 1a and thus choose to have your preference between f and g reversed at node 2. But notice that in this reasoning I had to assume that you anticipate that at node 1 you will choose in accordance with your current preferences (i.e., your preferences at node 0). An opponent

44

of transitivity can point out that he has already rejected the assumption that preferences should be constant; and indeed, it was precisely to try to get around that objection that we got involved with preferences about future preferences. So this move in fact gets us nowhere.

Thus the argument from modesty must be as I originally stated it. And what we have seen it to rest on is the following principle:

> In any decision tree, if you knew you would choose at future nodes in accordance with your current preferences, you would not prefer that your current preferences be different than they in fact are.

While I personally find this a plausible principle of rationality, I do not think that it is more compelling than transitivity and normality themselves. I have already indicated how McClennen would reject it.

2.2.4 Principles α and β

Sen (1971) shows that if a choice function C satisfies certain properties, then the choice function is normal, and preferences are transitive. This result can be regarded as providing yet another argument for transitivity and normality. It is the last argument I will consider.

One of the properties assumed by Sen is

Property α. *If S and T are sets of acts, and S is a subset of T, then any act f in both S and $C(T)$ is also in $C(S)$.*

That is: An act that is best in a given set of alternatives must also be best when the number of alternatives is reduced.

Another property needed for Sen's result is

Property β. *If S and T are sets of acts, and S is a subset of T, and if f and g are both in $C(S)$, then f is in $C(T)$ iff g is also in $C(T)$.*

That is: When the set of alternatives is enlarged, two acts that were initially best either both remain best, or both cease to be

45

best. Adding alternatives cannot make only one of the initially best options cease to be best.

I will say that a choice function C is *connected* if for any nonempty set of acts S, $C(S)$ is nonempty. Then the result proved by Sen is this: If C is a connected choice function satisfying properties α and β, then C is normal and binary preferences are transitive.

If we were to hold that rational persons must have connected choice functions, then this result would give us an argument for the rationality of transitivity and normality. But I have allowed that rational persons need not have connected choice functions. And Sen's result can fail when the choice function is not connected. That is, a person whose choice function is not connected can satisfy properties α and β, while violating transitivity or normality (or both).[4] Thus Sen's result as it stands does not give a sound argument for transitivity or normality. I now consider how this defect might be rectified.

Although a person can rationally have a choice function C that is not connected, it is plausible that C should satisfy this condition:

Extendability. *C is extendable to a connected choice function that can be held rationally.*

Extendability is satisfied just in case there is a rational connected choice function C^* such that, whenever $C(S) \neq \emptyset$, $C(S) = C^*(S)$.

Another plausible rationality requirement is

Closure. *If every rational connected extension C^* of C is such that $C^*(S) = T$, then $C(S) = T$.*

In other words, if the requirement of extendability fixes what $C(S)$ must be if it is nonempty, then $C(S)$ must now have that value. One might object to closure on the ground that it may take an unreasonable amount of thought to figure out that $C(S)$ can be given a nonempty value in only one way, if extendability is not to be violated; but I assume we are dealing

---

[4]For example, if $f \in S$, and $g \precsim f$ for all $g \in S$, but $C(S) = \emptyset$, then normality is violated, but there need be no violation of α or β.

with situations in which the options are important enough that doing the necessary thinking is worthwhile.

Now a modified version of Sen's result, which applies even to unconnected choice functions, is this: If a choice function satisfies principles α and β, plus extendability and closure, then normality and transitivity are satisfied. (I omit the proof.)

If the reader finds this a compelling argument for transitivity and normality, well and good. But my own judgment is that while I accept the premises of the argument, I don't find them more compelling than transitivity and normality themselves. The attractiveness of principles α and β seems to derive from the view of acts as ordered in a way that does not depend on what other acts are available, and this is just what transitivity and normality assert. Furthermore, the fact that we needed to appeal to two additional principles (extendability and closure) tends to undercut the force of the justification.

This completes my discussion of arguments for transitivity. I have contended that one of these arguments (the money pump argument) is fallacious. If somebody finds one or more of the other arguments compelling, I have no quarrel with that. But I have stated my own view that when the assumptions of these arguments are carefully scrutinized, the arguments do not succeed in adding to the initially high plausibility of transitivity and normality. This result can be seen as a compliment to the plausibility of transitivity and normality, rather than a strike against them. But if we have no good argument *for* transitivity or normality, it becomes all the more important to consider whether there is a good argument *against* them; to this I now turn.

2.3 OBJECTIONS TO TRANSITIVITY AND NORMALITY

Some writers on decision theory have explicitly rejected transitivity; a smaller number have also rejected normality. In this section, I will discuss some of the reasons that have been given for rejecting these conditions. The discussion will mostly concern transitivity, since this is the condition that has received most critical attention; but I will discuss one objection that is aimed at normality as well as transitivity (Levi's). I will argue

47

that the reasons offered for rejecting transitivity and normality have little, if any, force.

Recall that one possible objection to transitivity has already been dealt with in Section 1.5. We saw there that if lacking a view about the relative merits of two options were identified with indifference between those options, then transitivity could well be violated. I headed off this objection by insisting on a distinction between those two attitudes. If you lack a view about the relative merits of f and g, then the preference relation, as I interpret it, is not defined between these options.

The objections to transitivity that I will now discuss all purport to show that there can be good reasons to violate transitivity, even when preference is interpreted as I have proposed. I start with the least cogent objection and end with what I take to be the most serious challenge to transitivity.

2.3.1 Probabilistic prevalence

If a, b, and c are three quantities, it is possible that $p(a < b)$, $p(b < c)$, and $p(c < a)$ are all greater than $1/2$. Bar-Hillel and Margalit (1988) refer to this phenomenon as *probabilistic prevalence*. Blyth (1972), Packard (1982), Anand (1987), and Bar-Hillel and Margalit have all claimed that the phenomenon leads to intransitive preferences.[5]

Consider a concrete example, due essentially to Blyth. Let a, b, and c denote the times taken by three runners A, B, and C, in a race. Assume the probability distribution of a, b, and c is as follows. (Here times are in minutes (') and seconds (").)

$$p(a = 1'0'', \ b = 1'1'', \ c = 1'2'') \ = \ .3$$
$$p(a = 1'2'', \ b = 1'0'', \ c = 1'1'') \ = \ .3$$
$$p(a = 1'1'', \ b = 1'2'', \ c = 1'0'') \ = \ .4.$$

Then $p(a < b) = .7$, $p(b < c) = .6$, and $p(c < a) = .7$.

---

[5]Bar-Hillel and Margalit argue for intransitivity in choice patterns, which they distinguish from preference. But given the connection between preference and choice that I have set out in Section 1.4, their intransitivities in choice are intransitivities in preference. (Their discussion of the relation between preference and choice, in the final section of their paper, (a) conflates preferences with motives; and (b) erroneously claims that choice is extensional.)

Blyth does not explain how this phenomenon is supposed to lead to intransitive preferences. Bar-Hillel and Margalit fill out the argument; they say that "A is a better bet than B when these two are the competing runners, B is a better bet than C when these two are the competing runners, and C is a better bet than A when these two are the competing runners" (1988, p. 131). They claim we have here a case of strict intransitivity which is "perfectly justified and rational" (1988, p. 132).

But in fact, there is no violation of transitivity in this example. Let f_B^A be the act of betting that A will beat B; and similarly for the other pairs of runners. Assuming these bets are at even odds, then what we have in this example is that

$$f_B^A \succ f_A^B; \quad f_C^B \succ f_B^C; \quad f_A^C \succ f_C^A.$$

But this is a violation of transitivity only if $f_A^B = f_C^B$, $f_B^C = f_A^C$, and $f_C^A = f_B^A$. Now an act determines what consequences will be obtained in what states. Indeed, in Savage's (1954) presentation of decision theory, acts are identified with functions from the set of states to the set of consequences. Thus two acts can be said to be identical just in case they give the same consequences as each other in every state. One possible state is

$$a = 1'0'', \ b = 1'1'', \ c = 1'2''.$$

In this state f_A^B gives a loss, while f_C^B gives a profit. So f_A^B and f_C^B give different consequences in some states, and hence are not identical acts. The other required identities are similarly shown to be false.

2.3.2 Shifting focus

Arthur Burks writes:

A subject may prefer act a_1 to act a_2 on the basis of one criterion, a_2 to a_3 on the basis of another criterion, a_3 to a_1 on the basis of a third, and yet be unable to measure the relative strength of his desire for each act in terms of each criterion or to weigh the relative merits of the three criteria. For example, he may prefer a Volkswagen to a Ford on the basis of size, a Ford to a Plymouth on the basis of safety, a Plymouth to a Volkswagen on the basis of speed, and yet be unable to rank the criteria of size, safety, and speed. In this case it seems reasonable for his preferences to be intransitive. (1977, p. 211)

The person in Burks's car example regards smallness, safety, and speed as desirable features in a car. Smallness is a reason for favoring the Volkswagen over the Ford, but on the other hand, speed and safety are reasons to favor the Ford over the Volkswagen. But then, the fact that the Volkswagen is smaller than the Ford is no compelling reason to prefer the Volkswagen to the Ford overall. It may be said that the Volkswagen is *much* smaller than the Ford, while the differences in speed and safety are not so dramatic. But as against that, it can be said that speed and safety are two factors, while size is only one. So we have been given no good reason why the Volkswagen should be preferred to the Ford overall. Similarly, we have been given no good reason why the Ford should be preferred to the Plymouth, or the Plymouth to the Volkswagen. Hence, I do not see any good reason to think that in this case it is reasonable to have intransitive preferences.

Burks is careful to say that the person in his example is unable to decide on the relative importance of the three desiderata. But in that case, it seems that the person ought to lack a view about whether the Volkswagen is better than the Ford, not focus on one desideratum for this comparison and ignore the others. So this feature of Burks's case does not provide any support for the conclusion that preferences can reasonably be intransitive.

After the passage quoted, Burks goes on to say the reason for not ranking the three options in a single transitive ordering may be that it is hard to do. But if we are looking for an easy basis for choosing between the options, we could simply rank all three on the basis of, say, safety alone. There is no reason why one comparison should be made on the basis of one desideratum alone, while other comparisons are made by using different desiderata in isolation. Anyway, the case now seems to be becoming one in which the rationale for intransitivity is that the decision problem is not sufficiently important to justify the computational costs that might be involved in achieving transitivity; and I have conceded (in Section 1.3) that intransitivity need not be irrational under these conditions.

So while shifting focus in different comparisons can lead to intransitive preferences, I do not think we have been given any reason for doing that, and certainly not in decision problems of sufficient importance. Hence this observation does not undermine the view that transitivity is a requirement of rationality, when the preferences are relevant to a sufficiently important decision problem.[6]

It is well known that majority rule can produce intransitive pairwise rankings. For example, suppose Fred, Gary, and Helen rank order the options f, g and h as shown in Figure 2.6 (with 1 being the highest ranking). Then in a choice between f and g, a majority vote will select g; in a choice between g and h, a majority vote will select h; and in a choice between h and f, a majority vote will select f. This situation is sometimes referred to as the *Condorcet paradox*.

There have been several attempts to argue from the Condorcet paradox to the conclusion that intransitive preferences are sometimes reasonable. Perhaps the simplest such argument is one given by Bar-Hillel and Margalit (1988, p. 122). They suggest that a benevolent dictator might reasonably want to make whatever choices a majority of the subjects would vote for. In view of the Condorcet paradox, they immediately conclude that this dictator may have intransitive preferences.

The most natural interpretation of this scenario is that the dictator puts a value on doing what the majority would vote for. But if so, then there is no violation of transitivity. For example, suppose the dictator's subjects are Fred, Gary, and Helen, with rank orderings as in Figure 2.6. Then the consequence of choosing g over f for the dictator is not just whatever g gives but also the desired consequence of doing what the majority wants. Since an act must specify the consequences obtained in each state, and since consequences must be fully specific with regard to everything the decision maker cares about (Section 1.1), the act that the dictator chooses in this case cannot be identified

---

[6]R. I. G. Hughes (1980) also argues for intransitivity on the basis of shifting focus. Unlike Burks, Hughes fills out his example in such a way as to make the preferences compelling. But so filled out, the options in the various pairwise comparisons are not the same, and there is no violation of transitivity. His mistake is the same one involved in the argument from probabilistic prevalence.

	f	g	h
Fred	1	3	2
Gary	2	1	3
Helen	3	2	1

Figure 2.6: Preference rankings

with g. Let us call it instead g^+. Similarly, what choosing f would give in this case is not merely what f would give but also the undesirable consequence of not doing what the majority would vote for; we could call this option f^-. Continuing in this way, we find that the dictator's preferences are $f^- \prec g^+$, $g^- \prec h^+$, and $h^- \prec f^+$. This is not a violation of transitivity, because $f^- \neq f^+$, and similarly for g and h.

But perhaps it will be stipulated that this dictator does not put any intrinsic value on doing what the majority would vote for; instead the dictator cares only for the welfare of the subjects and on this basis prefers options that would be chosen by the majority. In that case, I would suggest that since the majority votes here produce intransitive pairwise rankings, the dictator has good reason to think that in this case the majority vote is a poor guide to what is best for the welfare of the subjects. Majority rule may be desirable because of such considerations as fairness and practicality, but apart from those considerations (which must be ignored to get a counterexample to transitivity), there is no reason to think that majority rule always maximizes the welfare of the subjects. Indeed, if the subjects are Fred, Gary, and Helen, and they are faced with the decision tree of Figure 2.4, a majority would at node 1 vote not to use majority rule at node 2. (The proof is essentially the same as the proof that strictly intransitive preferences are modest, given in Section 2.2.3.)

Another attempt to use Condorcet's paradox against transitivity considers a situation in which you must choose between options that differ in several different aspects that you care about. It is claimed that in such a case, you could reasonably prefer one option to another just in case it is superior in a

52

	Volkswagen	Plymouth	Ford
Size	1	2	3
Safety	2	3	1
Speed	3	1	2

Figure 2.7: Ranking of car qualities

majority of the relevant respects. And this rule leads to intransitive preferences.

For a concrete illustration, let us reconsider Burks's car example. Suppose that the ranking of the three cars with respect to size, safety, and speed is as in Figure 2.7. Applying majority rule to the aspects, we get that the Volkswagen is better than the Plymouth, the Plymouth is better than the Ford, and the Ford is better than the Volkswagen. (The opposite of the preferences Burks thought reasonable in this case!)

However, the fact that the Volkswagen is better than the Plymouth in two respects is not much of a reason to prefer the Volkswagen. The respect in which it is worse may be much more significant. If you have no opinion about the magnitude of the differences in the various aspects, then it seems to me more natural to be indifferent between these options, or else to lack a view about which of these options is best. So I see no great intuitive appeal in basing preferences on majority rule among attributes. Thus the fact that preferences formed in this way may be intransitive is not a good reason to deny that rational preferences must be transitive.

2.3.4 Levi

In a series of works (1974, 1980, 1986), Isaac Levi has insisted that rational persons need not have precise probabilities and utilities for all conceivable events and consequences. I have myself endorsed this view in Section 1.5. However, Levi's conception of what is involved in having indeterminate probabilities and utilities is different from mine. On the preference interpretation I have adopted, to have indeterminate probabilities or

53

utilities is to lack a view about what choices are best in some decision problems. Levi, on the other hand, provides rules that would determine, for any rational agent and any decision problem, which options are best in that decision problem. Thus Levi is committed to connectedness, in the sense in which I use that term.[7] And as Levi himself has emphasized (1986, ch. 6), his rules for making choices when probabilities and utilities are indeterminate violate both transitivity and normality. Here I will briefly review those rules and argue that there is no compelling reason to accept them.

Levi refers to the set of probability functions that represent the person's probability judgments as the person's *credal state*. And he refers to the set of utility functions that represent the person's value judgments as the person's *value structure*. When probabilities are indeterminate, the credal state will contain more than one probability function; and when utilities are indeterminate, the value structure will contain essentially different utility functions. An option that maximizes expected utility relative to some probability function in the credal state and some utility function in the value structure is said to be *E-admissible*. Levi holds that E-admissibility is a necessary, but not sufficient, condition for a choice to be rational. He suggests that considerations of security may be used to discriminate between the E-admissible options. Given some partition, the *security level* of an act is the minimum of the utilities of the act given each element of the partition. Levi leaves it up to the person to determine what partition will be used for assessing security. He says an option is *S-admissible* if its security level (relative to the agent's favorite partition) is at least as great as that of any other E-admissible option. Levi holds that a rational choice should be S-admissible. If there is more than one S-admissible act, then further tests may be used to distinguish between them. The acts that survive all tests are said by Levi to be "admissible"; using the terminology of Section 1.6, I would say these acts constitute the choice set.

---

[7]The concept of preference that I set out in Section 1.4 is what Levi (1986, p. 97) calls "basic revealed" preference; and that relation is connected for a rational person, according to Levi's decision theory.

54

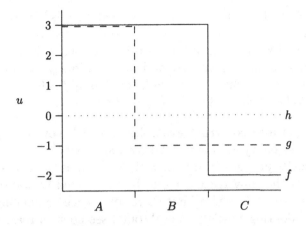

Figure 2.8: Example in which Levi's theory violates transitivity

To see that this proposal violates transitivity, let $\{A, B, C\}$ be a partition of the states of nature, and let acts f, g, and h be as follows:[8]

$$u(f) = 3 \text{ on } A \cup B, -2 \text{ on } C$$
$$u(g) = 3 \text{ on } A, -1 \text{ on } B \cup C$$
$$u(h) = 0 \text{ everywhere.}$$

These acts are represented in Figure 2.8. Suppose that $A \cup B$ has a determinate probability of $2/3$, but that the probabilities of A and B are each indeterminate with a lower bound of 0 and an upper bound of $2/3$. Suppose further that security is assessed relative to the partition $\{A, B, C\}$ and that no considerations beyond E-admissibility and security are invoked. Then Levi's theory requires $f \prec g$, $g \prec h$, and $h \prec f$, giving strictly intransitive preferences.

Proof. In a choice between f and g, both options are E-admissible, since $EU(f) = 4/3$ for every probability function in the credal state, while $EU(g)$ ranges from -1 to $5/3$. Since

---

[8]By "$u(f) = 3$ on $A \cup B$", I mean that for all states $x \in A \cup B$, $u[f(x)] = 3$. Similarly for the other cases.

55

the security level of f is -2, while that of g is -1, only g is S-admissible in a choice between f and g, whence $f \prec g$. In a choice between g and h, both options are again E-admissible but h has a security level of 0, and so h is uniquely S-admissible; hence $g \prec h$. In a choice between h and f, only f is E-admissible, since $EU(f) = 4/3 > 0 = EU(h)$; and so $h \prec f$.

One can also construct examples in which Levi's proposals violate normality, without violating transitivity.[9]

So Levi's proposal conflicts with both transitivity and normality. Is that any reason not to be committed to transitivity and normality? It would be if there were some compelling reason for adopting Levi's proposal; but I see no such reason. For one thing, Levi's proposal presupposes that a rational person must have a view about which acts are best in any decision problem; as I said in Section 1.5, this is implausible.[10]

But then what advice would I offer you, when you are faced with a decision problem, in which you lack a preference about what to choose? In some cases, I would give you specific advice based on information I have that you may lack; for example, if you have no preference about whether or not to take an umbrella, and I have reason to think it will rain, I will advise you to take an umbrella. But what if you and I have the same information? I still might accept norms that govern your case as I understand it, and give you advice based on these. For example, if you have no preference about whether or not to mutilate yourself for no reason, my advice would be not to do it, because I think this would be irrational. In other cases, the norms I accept would say that any option can be rationally chosen, in which case my advice would be that it doesn't matter what you choose. Finally, my norms might be like your preferences and

[9] One such example is provided by Cases 1, 2, and 4 of (Levi 1974), when P/S is between .4 and .5. Levi there cites this example as showing that his theory violates the "principle of independence of irrelevant alternatives."

[10] In a paper presented at the 1986 Philosophy of Science Association meetings (Maher 1986a), I pointed out that there are many alternatives to Levi's decision rule and claimed that Levi had given no reason for preferring his rule to these alternatives. Levi, who was in the audience, responded that his account fits choices that are often made in the Allais and Ellsberg problems. That response prompted the research reported in (Maher 1989) and (Maher and Kashima 1991), which shows that the choices Levi is referring to are not made for the reasons envisaged by his theory. (These choices will be discussed in Chapter 3.)

not extend to the decision problem at hand; then I would say I don't know what you should do. I think my advice, incomplete as it may be, is likely to be better than Levi's arbitrary rules.

2.3.5 Discrimination threshold

It is plausible that when the difference between two options is too small to be perceptually detectable, then a person would be indifferent between those options. But if this prima facie plausible thesis is accepted, then transitivity of preferences becomes unsupportable.

To illustrate this, suppose you prefer your coffee with a spoon of sugar in it.[11] You surely cannot distinguish between cups of coffee that differ by only one grain of sugar. If so, and if you are indifferent between options that are not perceptually distinguishable, then you violate transitivity.

Proof. Let n denote the number of grains in a spoonful of sugar, and let g_i denote the option of receiving a cup of coffee with i grains of sugar in it. By assumption, $g_i \sim g_{i+1}$, for all i. One application of transitivity gives $g_0 \sim g_2$. A second application gives $g_0 \sim g_3$. Continuing in this way, we get by $n-1$ applications of transitivity that $g_0 \sim g_n$. But you do not satisfy this condition; for you, $g_0 \prec g_n$.

If we are to maintain transitivity, and also that your preferences here are rational, it seems we must say that despite appearances, you do prefer a cup of coffee with $i+1$ grains of sugar to one with i grains, for some i between 0 and $n-1$. In defense of that option, it can be said that this preference need only be very slight – so slight that it would be outweighed by other small differences. For example, you needn't be prepared to pay as much as a penny more for the cup with the extra grain of sugar, or even to reach significantly further to obtain it. Also, there is a cost of calculation, which makes it uneconomic to worry much about very minor differences. Thus in practice, a very slight preference will be essentially indistinguishable from indifference. Ordinary language might well not distinguish between such a slight preference and true indifference; in ordinary

[11]This example is derived from (Luce 1956).

57

life, there would be little point in making the distinction. But when we philosophize about long sequences of options, each similar to the next but with very different endpoints, we can use this distinction between very slight preference and indifference to resolve an apparent contradiction between reasonable-seeming preferences and a commitment to transitivity.

That defense of transitivity could be criticized, as follows. The reason you want sugar in your coffee is presumably for its effect on the taste. But you cannot detect the difference between cups of coffee differing by only one grain of sugar. Hence the taste must be the same in each case; and since the taste is the only reason you care about having sugar in coffee, it follows that you must be indifferent between cups of coffee differing by only one grain of sugar. So there must be a violation of transitivity here, after all.

I will argue that this objection is fallacious. But before criticizing it, let me clarify one possible source of confusion. Suppose you are presented with two cups of coffee differing by only one grain of sugar and you have no idea which is which. Then surely you will be indifferent as to which one you choose. But it does not follow from this that you are indifferent between cups of coffee differing by only one grain of sugar. The situation is rather that each of the choices open to you is a gamble with a 50-50 chance of giving either a cup with i grains of sugar, or one with $i + 1$ grains, for some i. Analogy: Suppose you are presented with two black boxes, one containing a million dollars and the other containing shredded newspaper, so that you cannot tell which is which. Offered your choice between these boxes, you will probably be indifferent between the two options open to you; but you are not indifferent between getting a million dollars and getting shredded newspaper.

So the sort of case we need to focus on is one in which you not only know that the two cups of coffee differ by only one grain of sugar, but also are informed which one has the extra grain, although you could not distinguish which was which if you had not been told. And the question that concerns us is whether, given that your only reason for wanting sugar in your coffee is for its taste, it follows that you must be indifferent

58

between receiving either of these cups of coffee. I maintain that it does not follow. To fix ideas, suppose the following are true for direct comparisons of cups of coffee:

(i) You cannot distinguish by taste between a cup containing no sugar and one containing 10 grains of sugar.

(ii) You cannot distinguish by taste between a cup containing 10 grains of sugar and one containing 20 grains.

(iii) You *can* distinguish by taste between a cup containing no sugar and one containing 20 grains of sugar.

According to the argument of two paragraphs back, (i) implies that for you, the taste of coffee containing no sugar is the same as the taste of coffee containing 10 grains of sugar. However, this implication is demonstrably false in the case at hand. By (iii), you can taste the difference between cups of coffee containing 0 and 20 grains of sugar. If we replace the sugarless coffee by a cup containing 10 grains of sugar, you can no longer taste a difference with the 20-grain cup. Hence, you can by taste distinguish between the 0-grain and 10-grain cups; so these must taste different. The fact that you cannot distinguish them in a direct comparison therefore shows, not that they produce the same taste sensation, but rather that in a direct comparison you are unable to distinguish the taste sensations they produce. And given that the taste is different, you can prefer the coffee with 10 grains of sugar to the sugarless coffee, even though you cannot directly distinguish between these on the basis of taste, and even though taste is the only reason you care for sugar in your coffee. The fallacy in the argument of two paragraphs back is that it tacitly makes the false assumption that differences of sensation must always be detectable in direct comparisons.

Let us use the notation $a \succ\!\!\succ b$ to denote that a tastes detectably better than b in a direct comparison, and use $a \approx b$ to denote that a and b cannot be distinguished on the basis of taste in a direct comparison. Then it is plausible that a really tastes better than b just in case one of the following three conditions holds:

- $a \gg b$;
- $a \approx b$ and there exists c such that $a \approx c$ and $c \gg b$; or
- $a \approx b$ and there exists d such that $a \gg d$ and $d \approx b$.

If so, then someone who cares only about taste should have $a \succ b$ just in case one of these three conditions holds. Luce (1956) shows that, provided $\gg$ and $\approx$ satisfy reasonable conditions,[12] the preferences of such a person are transitive. So, contrary perhaps to first appearances, there is an attractive way of conforming to transitivity in the coffee example.

2.4 HUME AND MCCLENNEN

I have now considered arguments for and against transitivity and normality, and found both wanting. That leaves us where we started: Transitivity and normality are attractive normative principles endorsed by the vast majority. While violations are not rare, almost everyone feels they should correct those violations when they discover them. I think this is as good a case as we can hope to get for substantive normative principles; if further support is desired, it should be sought in successful applications of the principles. Thus I propose to proceed with a normative theory incorporating transitivity and normality.

The position I am taking here has been criticized by McClennen. He notes that there is not a consensus on principles like transitivity and normality, and holds that "Where there is no

---

[12]The conditions are that for all a, b, c, and d:

- Exactly one of $a \gg b$, $b \gg a$, or $a \approx b$ obtains;
- $a \approx a$;
- If $a \gg b$, $b \approx c$, and $c \gg d$, then $a \gg d$;
- If $a \gg b$, $b \gg c$, and $b \approx d$, then not both $a \approx d$ and $c \approx d$.

To forestall possible misunderstanding, let me make it clear that although I am using Luce's formal result, the interpretation I am putting on that result is different from Luce's. Luce accepted that intransitive indifferences were reasonable, and offered his result to show that utilities could be defined nonetheless, with indifference occurring when the difference in utility was smaller than some threshold. His abandonment of transitivity may be due to a failure to make the distinctions I have drawn in the two preceding paragraphs.

consensus, it is important to proceed with the greatest of caution. Otherwise, one ends up legislating for all about matters best left to individual discretion" (1990, p. 86). He proposes to follow Hume, and say that choice is irrational only if it is inconsistent with the chooser's ends. As McClennen interprets it, this seems to boil down to saying that the only principle of rationality is that you should not make choices that are certain to leave you worse off than some other alternative that was available. As the previous discussion of modesty will have indicated, a person who violates transitivity and normality need not violate this principle, provided preferences are revised over time in suitable ways. Thus McClennen rejects transitivity and normality as requirements of rationality.[13]

McClennen's conception of rationality is similar to the consistency conception discussed in Section 1.7. As I argued there, this is a notion of rationality that is not adequate for the sort of purposes that concern me. The principle that one should not accept sure losses is not a sufficient basis on which to build a normative theory of scientific method, and thus acceptance of McClennen's strictures would make such a theory impossible.

Let us then evaluate McClennen's argument for his weak conception of rationality. His premise is that we should not adopt normative principles that lack unanimous support. Why should we accept this premise? Perhaps unanimity on normative principles is a desirable thing, and one way to achieve this would be for everyone to weaken the norms they accept to the common denominator. But this would involve many of us abandoning normative principles we accept, and speaking for myself, I don't think this is a price worth paying. Thus I doubt that McClennen's premise itself enjoys anything like unanimous endorsement. Thus if we accept McClennen's premise, we should reject it. Hence we should reject it.

---

[13]Though this is not how McClennen himself puts the matter. He says that he accepts transitivity for a fixed set of options, and rejects only that the ordering must be unchanged by adding or removing options. But this separates preference from choice in a way that I think makes the notion of preference meaningless. For example, I can make no sense of what it would be to have $f \prec g \prec h$ if the options are f, g, and h, but $g \prec f$ if the options are f and g. As I use the term, binary preference relates to choices where the available options are binary. With preference so understood, McClennen rejects transitivity and normality.

I suggest that our reaction to differences of opinion over transitivity and normality should instead be to let "a hundred flowers blossom, and a hundred schools of thought contend" (Mao 1966, ch. 32). Since foundational arguments have been found inadequate to settle the issue either way, advocates of different positions should get to work developing theories based on their preferred principles. We can then use our judgments of the resulting theories to help decide between the principles. In later chapters of this book, I try to develop further the theory based on transitivity and normality by applying it to the acceptance of scientific theories and the delineation of scientific values. I encourage critics of transitivity and normality to tell us how they would deal with these issues.

3

Independence

Having argued for the legitimacy of assuming transitivity and normality, I now turn to the independence principle (defined in Section 1.3). Independence does not have the kind of popular endorsement that transitivity does. Nevertheless, I show that independence does follow from premises that are widely endorsed. The argument that I give differs at least subtly from those in the literature, and I explain why I regard my argument as superior. I also explain why arguments against independence strike me as uncompelling.

3.1 VIOLATIONS

In Chapter 2 we saw that most people endorse transitivity and normality. With independence, on the other hand, there is fairly strong prima facie evidence that many people reject it. Part of this evidence consists of decision problems in which a substantial proportion of people make choices that appear inconsistent with independence; there is also evidence that to many people the axiom itself does not seem compelling.

3.1.1 The Allais problems

Maurice Allais was one of the earliest critics of independence, and backed up his position by formulating decision problems in which many people have preferences that appear to violate independence. I will present these problems in the way Savage (1954, p. 103) formulated them, since this brings out the conflict with independence most clearly.

A ball is to be randomly drawn from an urn containing 100 balls, numbered from 1 to 100. There are two separate decision problems to consider, with options and outcomes as in Figure 3.1. Studies have consistently found that a substantial proportion

Problem A

	1	2–11	12–100
a_1	$1,000,000	$1,000,000	$1,000,000
a_2	$0	$5,000,000	$1,000,000

Problem B

	1	2–11	12–100
b_1	$1,000,000	$1,000,000	$0
b_2	$0	$5,000,000	$0

Figure 3.1: The Allais problems

of subjects choose a_1 in Problem A and b_2 in Problem B. Assuming that these choices reflect strict preferences, it follows that these subjects have $a_2 \prec a_1$ and $b_1 \prec b_2$ – preferences that are inconsistent with independence if the consequences are taken to be the monetary outcomes. Allais claimed that these were reasonable choices to make, so I will call them the *Allais choices*. (The corresponding strict preferences will be called the *Allais preferences*.) In versions of the Allais problems sometimes differing slightly from that given here, the proportion of subjects making the Allais choices has been found to be 46% (Allais 1979), 80% and 50% (Morrison 1967), 39% (MacCrimmon 1968), 30% (Moscowitz 1974), 35% to 66% (Slovic and Tversky 1974), 33% (MacCrimmon and Larsson 1979), 61% (Kahneman and Tversky 1979), and 38% (Maher 1989).

For all that I have said so far, it might be that the Allais choosers are committed to independence but fail to realize that their preferences violate that commitment. This hypothesis has been investigated, by presenting subjects with an argument against the Allais choices, based on independence; and an argument for those choices, conforming to Allais's reasoning. It is consistently found that most subjects rate Allais's

64

	1	2–11	12–100
a_1	$1,000,000	$1,000,000	$1,000,000
a_2	$0 + L	$5,000,000	$1,000,000
b_1	$1,000,000	$1,000,000	$0
b_2	$0	$5,000,000	$0

Figure 3.2: Modified consequences for the Allais problems

reasoning more compelling than that based on independence; and after reading these arguments, the proportion endorsing the Allais preferences is apt to increase slightly (MacCrimmon 1968, Slovic and Tversky 1974, MacCrimmon and Larsson 1979).

A number of writers (Morrison 1967, Raiffa 1968, Bell 1982, Eells 1982, Jeffrey 1987, Broome 1991) have suggested that those who make the Allais choices think it would be more unpleasant to have ball 1 drawn if a_2 was chosen than if b_2 was chosen. As Morrison (1967) puts it, if you choose a_2 and ball 1 is drawn, you may feel that you have lost $1 million rather than that your wealth remains unchanged; and you may not feel this way if you choose b_2 and ball 1 is drawn. But if this is so, then since consequences must specify everything you care about (Section 1.1), the consequence of choosing a_2 when ball 1 is drawn is not the same as the consequence of choosing b_2 when ball 1 is drawn. If this is right, then the consequences in the Allais problems would be more accurately represented as in Figure 3.2 (where L denotes the sense of having lost a million dollars). If the consequences are identified in this way, then the Allais preferences do not violate independence.

While the Allais preferences can be reconciled with independence in this way, I think it unacceptably ad hoc in the absence of some independent evidence that subjects do indeed think that if ball 1 will be drawn, they would be worse off choosing a_2 than choosing b_2. Jeffrey (1987, p. 234) claims it is clear from the subjects' explanations of their preference that this is indeed the case, but I know of no formal study which supports this claim. The studies that have asked subjects for reasons in the

65

Allais problems (MacCrimmon 1968, Slovic and Tversky 1974, MacCrimmon and Larsson 1979) have merely asked subjects to indicate agreement or disagreement with the independence principle and with Allais-type reasoning. A representative example of the latter is the following.

> In Problem A, I have a choice between $1 million for certain and a gamble where I might end up with nothing. Why gamble? The small probability of missing the chance of a lifetime to become rich seems very unattractive to me.
>
> In Problem B, there is a good chance that I will end up with nothing no matter what I do. The chance of getting $5 million is almost as good as getting $1 million so I might as well go for the $5 million and choose b_2 over b_1 (Slovic and Tversky 1974, with notation changed).

The studies find that subjects tend to endorse this Allais-type reasoning; but it does not follow from this that subjects think they would be worse off with a_2 and ball 1 than with b_2 and ball 1.

What is needed is a study designed to test whether subjects do think they would be worse off with a_2 and ball 1 than with b_2 and ball 1. Until such a study is conducted, I think we cannot say with much confidence whether those with the Allais preferences are violating independence. But the fact that subjects rate Allais-type reasoning as more attractive than independence is reason to doubt that they are committed to independence.

3.1.2 The Ellsberg problems

Daniel Ellsberg (1961) also formulated decision problems in which it seems that many people have preferences that violate independence. One pair of decision problems, in which the apparent conflict with independence is most direct, is as follows.

A ball is to be randomly drawn from an urn containing 90 balls. You are informed that 30 of these balls are red and the remainder are either black or yellow, with the proportion of black to yellow balls being unknown. The two decision problems have the options shown in Figure 3.3. Ellsberg thought it reasonable to choose c_1 in Problem C, and d_2 in Problem D. His rationale

66

Problem C

	Red	Black	Yellow
c_1	$100	$0	$0
c_2	$0	$100	$0

Problem D

	Red	Black	Yellow
d_1	$100	$0	$100
d_2	$0	$100	$100

Figure 3.3: The Ellsberg problems

for choosing c_1 was that with it there is a known chance of $1/3$ of winning \$100, while with c_2 the chance of winning is unknown. His rationale for choosing d_2 was similar: with d_2 there is a known chance of $2/3$ of winning, while with d_1 all that is known is that the chance of winning is at least $1/3$. I will call these choices the *Ellsberg choices* and will call the corresponding strict preferences the *Ellsberg preferences*. These preferences violate independence, provided the consequences can be taken to be the monetary outcomes.

In studies where subjects were presented with Problems C and D, the proportion selecting the Ellsberg choices has been found to be 66% to 80% (Slovic and Tversky 1974), 58% (MacCrimmon and Larsson 1979), 75% (Maher 1989), and 60% (Maher and Kashima 1991). Assuming that these choices reflect strict preferences,[1] it follows that most subjects have the Ellsberg preferences.[2]

---

[1] For evidence confirming this assumption, see the data on Problem F in (Maher 1989) and (Maher and Kashima 1991).

[2] However, in two studies conducted at La Trobe University, Kashima and I (1992) found the proportion of Ellsberg choosers to be 34% and 25%. There were 131 subjects in the first study and 68 in the second, so these proportions differ significantly from those previously reported. There may be a cultural effect; La Trobe University is located in Melbourne, Australia, where gambling is part of the culture.

	Red	Black	Yellow
c_1	$100	$0	$0
c_2	$0 + F	$100 + F	$0 + F
d_1	$100 + F	$0 + F	$100 + F
d_2	$0	$100	$100

Figure 3.4: Modified consequences for the Ellsberg problems

Could it be that the Ellsberg choosers are really committed to independence and simply make a mistake in these problems? The evidence suggests otherwise. In some studies, subjects have been presented with an argument for making choices conforming with independence and an Ellsberg-style argument for the Ellsberg choices (Slovic and Tversky 1974, MacCrimmon and Larsson 1979); the result is that subjects who make the Ellsberg choices tend to rate the Ellsberg-style argument higher than the argument based on independence. A high level of agreement with Ellsberg-style reasoning was also reported in Maher and Kashima (1991).

A number of writers have suggested that persons who have the Ellsberg preferences put an intrinsic value on knowing chances (Eells 1982, Skyrms 1984, Jeffrey 1987, Broome 1991). As Skyrms (1984, pp. 33f.) puts it, ignorance of chances might induce the unpleasant sensation of "butterflies in the stomach." If this is right, then the consequences in Problems D and E would be more correctly represented as in Figure 3.4 (where F = butterflies in the stomach). If something like this is the correct representation of the consequences, then the Ellsberg preferences do not violate independence.

Like the similar move with the Allais problems, this way of making the Ellsberg choosers consistent with independence is unacceptably ad hoc in the absence of some independent evidence that Ellsberg choosers do indeed put an intrinsic value on knowing chances. The mere fact that these subjects tend to prefer gambles where the chances are known does not show that they put an intrinsic value on knowing chances. Analogy: People

68

generally prefer gambles in which the probability of winning is high, but Bayesians do not infer from this that people generally put an intrinsic value on having a high probability of winning. Rather, the high probability of winning is taken to be a means to the intrinsically desired end of actually winning. Similarly, subjects who make the Ellsberg choices might do so because they think this is a good means to the intrinsically desired end of receiving $100, not because they put any intrinsic value on knowing objective chances.

Yoshihisa Kashima and I conducted a study to ascertain whether Ellsberg choosers put an intrinsic value on knowing chances. The study was conducted with undergraduate students at La Trobe University. They were presented with a version of Problems C and D, with the $100 increased to $1,000. Seventeen of our subjects made the Ellsberg choices, and these were presented with the following reasons for those choices. Some were offered these reasons with respect to Problem C, others with respect to Problem D.

1. If I choose c_2 (d_1) I don't know the probability of winning, and that will make me feel nervous. If I choose c_1 (d_2), I know the probability of winning, and I'll feel more comfortable with that. So I choose c_1 (d_2) to avoid the feeling of anxiousness that comes with not knowing the probability of winning.
2. My only concern is to try to win the $1,000; feelings of anxiety don't bother me. I choose c_1 (d_2) because I think it makes sense to choose a known chance of achieving my goal, instead of an unknown chance of achieving it.
3. I agree with both reasons 1 and 2 above for choosing c_1 (d_2). That is, I choose c_1 (d_2) both to avoid the feeling of anxiety that would come with not knowing the probability of winning, and also because choosing a known chance of winning is a good way to win.

Subjects were asked to indicate to what extent they agreed with each of these three arguments. The result was that they agreed most with the second reason. This tends to disconfirm the hypothesis that Ellsberg choosers put an intrinsic value on knowing chances.

So the data on the Ellsberg choices, like that on the Allais choices, provide reason to think that many people are not committed to independence. However, I will argue that these people are making a mistake.

3.2 ARGUMENTS FOR INDEPENDENCE

3.2.1 Synchronic separability

I begin by presenting an argument that the Allais preferences are a mistake. This argument is due in its essentials to Markowitz (1959, pp. 221–4) and Raiffa (1968, p. 82f.), though my presentation will differ slightly from either of theirs.

Suppose that, as in the version of the Allais problems considered above, we have an urn containing balls numbered from 1 to 100. But now, before you have to make any choice, the experimenter draws a ball from the urn; if the number on the ball is 12 or greater, you receive either $1 million or nothing, depending on which version of the experiment is being conducted. If the number on the ball is 11 or less, the experimenter does not reveal what the number is, and gives you your choice between the following options:

f: $1 million.
g: $5 million if the number on the ball is not 1, and nothing if it is 1.

The situation is diagrammed in Figure 3.5. I follow the standard convention, in which boxes represent nodes at which a choice is made, and circles represent nodes at which "nature" or "chance" determines the branch to be taken.

Now suppose you could specify in advance what choice you would make, should a ball numbered 11 or less be drawn. That is to say, you can now tell the experimenter whether you want f or g should a ball numbered 11 or less be drawn, and the experimenter will take that to be your choice, should the drawn ball be numbered 11 or less. What choice do you want to specify in tree (a) of Figure 3.5? Is this the same choice you want to specify in tree (b)?

Markowitz and Raiffa report that most people specify the same choice for both trees. Markowitz reports that the invariable

70

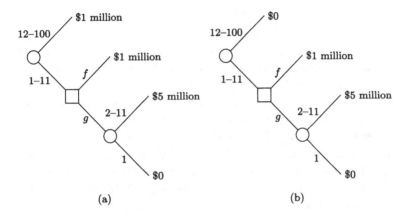

Figure 3.5: Sequential version of the Allais problems

choice was f – that is, the million dollars for sure. In any case, we have that for almost everyone, either

(i) Specifying f in advance is weakly preferred to specifying g in advance in both trees (a) and (b); or

(ii) Specifying g in advance is weakly preferred to specifying f in advance in both trees (a) and (b).

Furthermore, it seems uncontroversial that the consequences a person values are not changed by representing the options in a tabular or tree form. But then, specifying f in advance in tree (a) of Figure 3.5 is the same act as choosing b_1 in the Allais problems, while specifying g in advance in that tree is the same as choosing b_2. Hence (i) is inconsistent with the Allais preference $b_1 \prec b_2$. Similarly, (ii) is inconsistent with the Allais preference $a_2 \prec a_1$.

The data reported by Markowitz and Raiffa were obtained in informal discussion; but a formal study by Kahneman and Tversky (1979) points to the same conclusion. Kahneman and Tversky used a version of the Allais problems with different probabilities and prizes, and found that the majority of their subjects made Allais-type choices. However, when subjects were presented with Kahneman and Tversky's analog of tree (b) and asked what choice they would specify in advance, 78 percent

71

specified Kahneman and Tversky's analog of f, a choice inconsistent with the Allais preferences.[3]

McClennen (1983, n. 5) attempts to resist this conclusion by claiming the instructions that Markowitz, and Kahneman and Tversky, gave their subjects were ambiguous. He claims that when it is said the choice must be specified in advance, it is unclear whether one is being asked to predict one's later choice, or whether one is being asked to now make the choice. However, the instructions that Markowitz as well as Kahneman and Tversky report giving seem unambiguous to me. Markowitz's formulation of the problem (modified slightly to fit my presentation above) is

Suppose you had to commit yourself in advance. Suppose that, before the ball is drawn, you had to say "I will take f if the choice arises" or "I will take g if the choice arises." To which would you commit yourself?

Kahneman and Tversky refer to the chance and choice nodes in their version of the problem as the *first* and *second stages*; and they specify that

Your choice must be made before the game starts, i.e., before the outcome of the first stage is known.

This leaves no room for reasonable doubt that subjects are being asked to make the choice in advance.

We have been considering an argument against the Allais preferences. Kashima and I (1992) show that parallel points can be made about the Ellsberg preferences. That is to say, the Ellsberg preferences can be presented in a tree format, and when they are so presented most Ellsberg choosers cease to have the Ellsberg preferences.

This line of argument can be generalized to give an argument against *all* violations of independence, as I will now show.

I will say that you *currently prefer* choosing f rather than g at a choice node iff you now prefer specifying that you choose f were you to reach that node, rather than specifying that you choose g. The preference here can be understood as either strict

---

[3]The decision problems mentioned here are problems 3, 4, and 10 in (Kahneman and Tversky 1979).

72

or weak. So the data reported above can be expressed by saying that most people currently prefer that they choose f rather than g at both choice nodes in Figure 3.5. This data suggests that most people accept the following principle as normatively correct:

Synchronic separability. *Your current preferences regarding how you choose at a choice node do not depend on what would happen if you do not reach that choice node.*

For example, the difference between the choice nodes in trees (a) and (b) of Figure 3.5 consists solely in whether you would or would not have received a million dollars if you had not reached that node, and most people think this should not affect preferences regarding the choice at the node.

The term *separability* here denotes that the part of the decision tree prior to the choice node can be ignored in considering what choice is best at the node. I add the adjective *synchronic* because the preferences at issue are your current preferences only. (A principle of diachronic separability will be discussed in Section 3.2.2.)

Synchronic separability entails independence.

Proof. Suppose $f = f'$ on A, $g = g'$ on A, $f = g$ on $\bar{A}$, $f' = g'$ on $\bar{A}$, and $f \precsim g$. What needs to be shown is that if synchronic separability holds, it follows that $f' \precsim g'$.

In tree (a) of Figure 3.6, specifying that you choose g, should you reach the choice node, is equivalent to now choosing g over f. So since $f \precsim g$, you must now weakly prefer specifying that you choose g, as opposed to f. And this is the same as saying that you now weakly prefer that you choose g, should you reach the choice node in tree (a). By synchronic separability, it follows that you now have the same preference about your choice at the choice node in tree (b). So you now weakly prefer specifying that you choose g at the choice node in tree (b), as opposed to choosing f. But specifying g in advance in tree (b) is the same as choosing g' over f'. Hence you must have $f' \precsim g'$.

Thus we have an argument for independence, from premises that appear to be very widely endorsed.

73

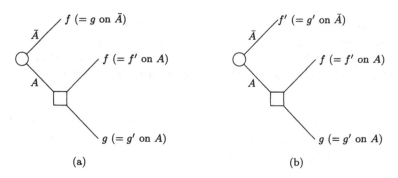

Figure 3.6: Trees to show synchronic separability entails independence

There are a number of arguments for independence in the literature, and the argument I have just given is closely related to many of them. However, so far as I know, the argument as I have given it is also at least subtly different from the arguments in the literature. The differences are for a reason: I think the argument as I have given it is superior to the arguments in the literature. So, at least for afficionados of this literature, I will explain why I think the argument from synchronic separability is superior to some of the arguments familiar in the literature. Since this discussion will not add any stronger argument for independence than the one just given, many readers may want to skip it and turn directly to the consideration of the arguments *against* independence, in Section 3.3.

3.2.2 Diachronic separability and rigidity

The argument of Markowitz and Raiffa, against the Allais preferences, has not usually been seen as appealing to synchronic separability. Instead, it has been seen as appealing to two principles, one of which is:

Diachronic separability. *The preferences you would have were you to reach a choice node do not depend on what would have happened, had you not reached that node.*

Applied to the sequential version of the Allais problems, this says that if a ball numbered 1–11 is drawn, the preference you

74

would then have will be the same whether you are in tree (a) or (b) of Figure 3.5. It does not say anything about your current preferences regarding your choice at that point. Hence the name *diachronic* separability.

Diachronic separability is not by itself inconsistent with the Allais preferences. One needs in addition the principle of

Rigidity. *Your current preferences regarding how you choose at a choice node are the same as the preferences you would have were you to reach that choice node.*

If rigidity is assumed, then diachronic and synchronic separability become equivalent. So rigidity and diachronic separability together entail synchronic separability, and hence entail independence. However, this implication is not a good reason to conclude that independence is a requirement of rationality. I gave some reasons for rejecting rigidity in Section 2.2.2; I will give additional reasons here.

One way preferences might rationally change, as you move through a decision tree, is that you acquired new information along the way. One might attempt to preclude this by stipulating that no relevant new evidence is acquired. But you might still use the time interval to reconsider your preferences, and may have some insight that leads to a rational revision of those preferences, without acquiring any new evidence. To preclude this, one would need to stipulate that a rational person is logically omniscient, so that nothing further can be learned by thought alone. However, since we are plainly not such creatures, this would be to concede that the argument for independence does not apply to us.

Another way preferences might rationally change is due to selective discounting of the future. It is usual in economics to suppose that future benefits are discounted according to how far in the future they lie. This in itself does not seem irrational. Nor would it be irrational to use different discounting schedules for different things. For example, future pains might be discounted more or less strongly than future pleasures. But then, if some possible future consequences contain different mixtures of pleasure and pain, a person's preferences between those consequences will change as the time of obtaining them approaches.

75

In Section 2.2.2, I indicated a way that any objection to rigidity can be met. We can stipulate that our criterion of personal identity is such that persons cease to be self-identical whenever they change their preferences. This makes rigidity not only true but tautologous. If we take this line, then diachronic separability has the same content as synchronic separability, and the present argument for independence is just a misleading way of giving the same argument that I gave in the preceding subsection.

3.2.3   The sure-thing principle

Savage (1954, pp. 21–3) says that independence (which he calls "P2") is suggested by what he calls the *sure-thing principle*.[4] Savage introduces this principle with the following example.

A businessman contemplates buying a certain piece of property. He considers the outcome of the next presidential election relevant to the attractiveness of the purchase. So, to clarify the matter for himself, he asks whether he would buy if he knew that the Republican candidate were going to win, and decides that he would do so. Similarly, he considers whether he would buy if he knew that the Democratic candidate were going to win, and again finds that he would do so. Seeing that he would buy in either event, he decides that he should buy, even though he does not know which event obtains.

This businessman is applying the sure-thing principle. Savage's "relatively formal" statement of the sure-thing principle is essentially as follows.

The sure-thing principle. *If you would have $f \precsim g$ were you to learn A and would also have $f \precsim g$ were you to learn $\bar{A}$, then you should now have $f \precsim g$. Moreover (provided you do not regard A as virtually impossible) if you would have $f \prec g$ were you to learn A and would also have $f \precsim g$ were you to learn $\bar{A}$, then you should now have $f \prec g$.*

---

[4] As I noted in Chapter 1, many authors have identified Savage's sure-thing principle with his P2 (= independence). Ellsberg's 1961 paper contains the earliest example of this identification that I have noticed. Jeffrey (1983) considers Savage's discussion of the sure-thing principle in some detail but makes the same identification. I think this identification fits poorly what Savage says. Also, the possibility of arguing from (what I call) the sure-thing principle to (what I call) independence makes it desirable to have different names for these principles.

76

The name of the principle comes from the idea that, under the stated conditions, it is a sure thing that g is at least as good as f. Savage says that "except possibly for the assumption of simple ordering [i.e., transitivity and connectedness], I know of no other extralogical principle governing decisions that finds such ready acceptance."

The phrase "virtually impossible," which appears in the sure-thing principle, is vague. But later Savage gives it a precise meaning: You regard A as virtually impossible just in case you are indifferent between all acts that differ only on A (1954, p. 24). Events that you regard as virtually impossible, in this sense, are also said to be *null*.

The "learning" referred to in the sure-thing principle must be understood in such a way that it satisfies this condition:

($*$) If you were to learn A, then you would be indifferent between acts that do not differ on A.

Another way of putting this would be to say that if you were to learn A, then $\bar{A}$ would be null.

With these understandings, the sure-thing principle entails independence.

Proof. Suppose $f = f'$ on A, $g = g'$ on A, $f = g$ on $\bar{A}$, $f' = g'$ on $\bar{A}$, and $f \precsim g$. Case (i): A is not null. For any preference relation R, let $A \to R$ denote that you would have R were you to learn A. By ($*$), $\bar{A} \to f \sim g$. If $A \to f \succ g$, we would then have by the sure-thing principle that $f \succ g$, which is not the case; hence $A \to f \precsim g$. Since $f = f'$ and $g = g'$ on A, it follows (by ($*$) and transitivity) that $A \to f' \precsim g'$. Since $f' = g'$ on $\bar{A}$, we also have $\bar{A} \to f' \sim g'$. So by the sure-thing principle, $f' \precsim g'$, as independence requires. Case (ii): A is null. Then since $f' = g'$ on $\bar{A}$, it follows from the definition of a null event that $f' \sim g'$, and so again $f' \precsim g'$.

Why is this not a better argument than the one from synchronic separability? Well, first of all, note that the sure-thing principle tacitly assumes that there is a fact about what preference you would have between f and g, were you to learn A. But if you violate diachronic separability, this need not be the case;

the preference you would have after learning A might depend on what you would have got if $\bar{A}$ had obtained. Thus the sure-thing principle, as stated by Savage, presupposes diachronic separability. Furthermore, this presupposition plays an essential role in the proof that the sure-thing principle entails independence.

Second, I note that the sure-thing principle entails rigidity. For suppose you violate rigidity, because you will have $f \precsim g$ in the future, though you now have $g \prec f$. Let A be any proposition whose truth value you will not learn until after your preference has changed. Then if you were to learn A or $\bar{A}$, you would have $f \precsim g$; so by the sure-thing principle, you should now have $f \precsim g$. Hence you violate the sure-thing principle.

Thus the argument from the sure-thing principle is not essentially different from the argument from diachronic separability and rigidity. And so it is open to the same objection: Rationality does not preclude changes of preferences over time.

Furthermore, empirical studies do not bear out Savage's claim that "except possibly for the assumption of simple ordering ... no other extralogical principle governing decisions ... finds such ready acceptance" as the sure-thing principle. MacCrimmon and Larsson (1979, sec. 6) found that subjects on average rated their agreement with the first part of the principle, on a scale from 0 to 10, as 6.7. This is not inordinately high; subjects indicated higher agreement with some principles of choice that are inconsistent with the sure-thing principle.

After this disparaging of the argument from the sure-thing principle, I should say that I think there is something right about the principle. I will use the notation $f \precsim_A g$ to denote that if you could now specify how you would choose were you to learn A, you weakly prefer specifying that you choose g. (Relations like $f \prec_A g$ and $f \sim_A g$ are defined similarly.) What I think is right is the

Synchronic sure-thing principle. *If $f \precsim_A g$ and $f \precsim_{\bar{A}} g$, then $f \precsim g$. Moreover, if A is not null, and if $f \prec_A g$, then $f \prec g$.*

As the name indicates, this is a synchronic analog of Savage's sure-thing principle. While I would endorse this principle, it is not much use in arguing for independence. For the principle

78

presupposes synchronic separability, and we have seen that synchronic separability by itself entails independence.

3.2.4 The value of information

A principle with considerable intuitive appeal is that it cannot be undesirable to acquire cost-free information. This principle follows from the principle of maximizing expected utility, provided suitable conditions are satisfied. (The "suitable conditions" are stated in Section 5.1.2.) Wakker (1988) suggests that independence follows from this attractive principle.[5]

For present purposes, it will be sufficient to present an example of how this argument is supposed to work. Suppose then that you have the Allais preferences but that (as most people apparently do) you would prefer f to g were you at either choice node in Figure 3.5. Now I present you with the following choice: You can either specify your choice in tree (b) of Figure 3.5 in advance, or else you can wait and see whether or not the ball is numbered between 1 and 11, and if it is, you then get to make your choice. This decision problem is represented in Figure 3.7.

Specifying f in advance is equivalent to choosing b_1 (from Problem B, Section 3.1.1), whereas specifying g in advance is equivalent to choosing b_2. Thus at node 2, your options are effectively between b_1 and b_2. Since you have the Allais preference $b_1 \prec b_2$, you would therefore choose g at node 2. Thus deciding to choose now is equivalent to choosing to specify g in advance, which is equivalent to choosing b_2.

On the other hand, we have assumed that you would prefer f to g at either choice node in Figure 3.5; hence[6] you would prefer f to g at node 3 in Figure 3.7. So if you postpone your decision until after you learn whether the ball is numbered 1–11, you will get $0 if it is not and $1 million if it is. This is

---

[5] Wakker presents an argument that certain violations of Samuelson's strong independence axiom entail that cost-free information is undesirable. But he says that the argument can be reformulated to apply to the "sure-thing principle" in place of strong independence. I guess that by the "sure-thing principle" he means what I am calling independence, and this is the argument I will consider here. But Wakker's argument is also easily adapted to give an argument for the sure-thing principle of Section 3.2.3.

[6] This isn't a logical entailment. It actually depends on an application of diachronic separability. Another way in which Wakker's argument presupposes diachronic separability will be noted shortly.

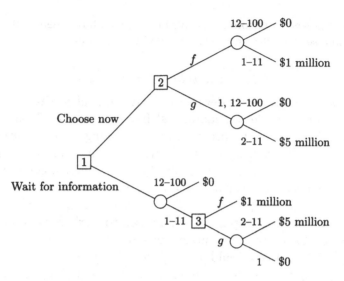

Figure 3.7: Decision of whether to acquire information

equivalent to getting b_1. Because you have the Allais preference $b_1 \prec b_2$, you therefore prefer to make your choice before getting the information. Thus you violate the principle on the value of cost-free information.

We have been assuming that you would have $g \prec f$ at both choice nodes in Figure 3.5. If instead you would have $f \prec g$ at both choice nodes, the argument is easily reformulated, using tree (a) of Figure 3.5 in place of tree (b) of Figure 3.5. But the argument does not work at all if you would have $g \prec f$ at the choice node in tree (a) but $f \prec g$ at the choice node in tree (b). Thus Wakker's argument tacitly assumes that diachronic separability holds (or at least that it is not violated in this way).

I turn now to what I take to be the main weakness in Wakker's argument. At node 1 in Figure 3.7, we are supposing you know you would choose f were you at node 3. But we have also seen that if you could specify your choice in advance, you would specify that you choose g. Thus you know that if you get the information, you will make a choice that you do not now think optimal for that situation. And it is really this that causes you not to want to acquire the information. But when your future choices will not be made in the way you now think would be

80

optimal, it is no longer intuitively plausible that cost-free information is not undesirable, nor does expected utility theory imply this result. Consequently, the fact that you would want to avoid cost-free information under these circumstances is not a reason to say you are irrational.

One might try answering this objection by saying that the real indicator of irrationality is that you know you would under certain circumstances choose in ways you now think suboptimal. But this is to assume a kind of rigidity condition, and is open to the same sort of objections as I have already raised.

To avoid these difficulties, we might try framing Wakker's argument in terms of *current* preferences. I would agree that if you know your future choices will be made in ways you now think optimal, then you should not want to avoid cost-free information. But recall that Wakker's argument also assumed diachronic separability; if we shift to current preferences, that assumption becomes synchronic separability. And we have seen that synchronic separability by itself suffices to derive independence. Hence the appeal to the value of information becomes redundant.

3.2.5 Other arguments

I have now discussed three alternatives to my argument from synchronic separability, and a common pattern has emerged: These alternative arguments all assume diachronic separability, which I think is no more attractive than synchronic separability. In addition, these arguments all tacitly assume rigidity, or something comparable, which is not a requirement of rationality.

There are other arguments for independence that I have not discussed. These include the arguments of Hammond (1988a,b) and Seidenfeld (1988). I will not discuss these arguments here, because what I would end up saying about them is essentially the same as what I have said about the last three arguments discussed: They assume principles akin to diachronic separability and rigidity.

So I conclude that the argument from synchronic separability is the best argument for independence that we have.

81

Fortunately, that argument is a good one; it derives independence from premises that most people accept.

3.3 OBJECTIONS TO INDEPENDENCE

There are able decision theorists who reject independence. It is time to look at what reasons they have to support this position. This can be brief, because the main points have already been covered in Section 3.1.

Allais, Ellsberg, and others have claimed that many people, who seem quite reasonable, do choose in a way that violates independence, and that this is a reason to think that a rational person need not satisfy independence. But as we have seen, the putative violations may not be genuine violations, because they may disappear when consequences are made sufficiently specific to include everything the person values in the outcome. Second, even if it can be shown that those with the Allais and Ellsberg preferences really are violating independence, there may be good reason to call them irrational; for they probably accept the premises of the argument from synchronic separability, in which case it is undeniable that they are making a mistake.

Allais thought there was a good reason to have the Allais preferences, namely that when an option has a sure payoff, the certainty itself provides a reason to choose that option, over and above what the payoff itself is. But if certainty has an intrinsic value, it needs to be included in the specification of the consequences; and once that is done, the Allais preferences do not violate independence. On the other hand, if certainty does not have an intrinsic value, Allais's argument in favor of the Allais preferences ceases to be compelling.

The situation is similar with Ellsberg. He maintained that there was a good reason to have the Ellsberg preferences, namely that when an option involves known chances, that fact itself provides a reason to choose the option. If knowing chances has intrinsic value, it needs to be included in the specification of the consequences; and once that is done, the Ellsberg preferences are consistent with independence. And if knowing chances does not have intrinsic value, Ellsberg's argument in favor of the Ellsberg preferences ceases to be compelling.

82

3.4 CONCLUSION

I have tried to show that there is no compelling argument against independence, and there is an argument for it from the attractive principle of synchronic separability. Of course, those decision theorists who oppose independence will respond by rejecting synchronic separability. I do not have further arguments to prove they are wrong, but they also lack an argument to prove their position. McClennen would claim that we should not adopt principles of rationality that do not enjoy unanimous endorsement, but I have already criticized that position in Section 2.4.

So while I do not claim to be able to bring opponents of independence to their knees, I do claim that the view that independence is a requirement of rationality is as reasonable as any contrary view. If the debate is to be taken further, I suggest that the most productive way to do it is for the different sides to develop theories based on their favored principles, so that we can judge the principles by their fruits. In this spirit, I now turn to the development of a theory based on the assumption that independence, along with transitivity and normality, is a requirement of rationality.

4

Subjective probability in science

4.1 CONFIRMATION

The preceding three chapters have been concerned with the Bayesian theory of rational preference. I turn now to a topic that may at first sight appear to be quite remote from those concerns, namely the question of the relation between scientific theories and the evidence for or against them. But as will quickly become apparent, these topics are intimately connected.

All scientific theories go beyond the evidence that is adduced in support of them; hence the truth of the theories does not follow by logic from that evidence. Even if the evidence is correct, it always remains possible that the theory is false. Nor is this a merely idle possibility. Newtonian mechanics was as well supported by evidence as any scientific theory is likely to be, but nevertheless proved to be not the literal truth. This fact naturally leads one to ask why it is that some evidence, which does not guarantee the truth of a theory, is nevertheless regarded as supporting, or confirming, the truth of the theory.

Bayesian philosophers of science hold that scientists have subjective probabilities for scientific theories. Evidence that confirms a theory is then taken to be evidence that increases the scientist's subjective probability for the theory. The question posed in the preceding paragraph is thus construed, from a Bayesian perspective, as a question about why some evidence increases the subjective probability of a theory.

Bayesian answers to this question make use of a principle of *conditionalization*, which we can state as follows.[1]

---

[1] Here '$\cdot$' represents an arbitrary event or proposition, and $p(\cdot|E)$ denotes probability conditioned on E. According to the usual definition, $p(A|E) = p(AE)/p(E)$, and is defined only if $p(E) \neq 0$.

Conditionalization. *If your current probability function is p, and if q is the probability function you would have if you learned E and nothing else, then q(·) should be identical to p(·|E).*

In Chapter 5 I will show that this principle follows from the principle of maximizing expected utility, under some reasonably mild additional assumptions.

The principle of conditionalization, together with the Bayesian understanding of what confirmation is, entails that evidence E confirms hypothesis H just in case $p(H|E) > p(H)$. Similarly, E disconfirms H just in case $p(H|E) < p(H)$.

Bayesian philosophers of science have shown that this rather modest-looking account of confirmation can explain many methodological intuitions about when, and how strongly, evidence confirms a hypothesis.

For a basic illustration of this, suppose that a hypothesis H, together with some background information in which we are highly confident, entails some testable proposition E. Suppose also that every plausible alternative hypothesis, together with background information, entails that E is false. Our intuitions are that in this case, the verification of E would confirm H. On the Bayesian account of confirmation, this intuition can be explained, as follows. The elementary theorem of probability known as Bayes' theorem entails that

$$p(H|E) = \frac{p(E|H)}{p(E|H)p(H) + p(E|\bar{H})p(\bar{H})}p(H).$$

Since $p(H) + p(\bar{H}) = 1$, we can rewrite this as

$$p(H|E) = \frac{p(E|H)}{p(E|H) - [p(E|H) - p(E|\bar{H})]p(\bar{H})}p(H).$$

Because E follows from H together with background information of which we are confident, $p(E|H)$ is high. And because every plausible alternative hypothesis predicts that E is false, $p(E|\bar{H})$ is low. Thus $p(E|H) - p(E|\bar{H}) > 0$, and so, assuming $p(\bar{H}) > 0$, we have

$$p(H|E) > \frac{p(E|H)}{p(E|H)}p(H) = p(H),$$

which is the desired result.

For a less trivial example, consider the following two scenarios.

(i) A hypothesis H is advanced, and H entails the empirical proposition E. Later it is discovered that E is indeed true.

(ii) E is discovered to be true, and then a hypothesis H is proposed to explain why E is true.

Many scientists and philosophers of science have maintained that, other things being equal, H would be better confirmed in situation (i) than in situation (ii). I call this view *the predictivist thesis*. Among those who have endorsed this thesis are Bacon (1620, book I, cvi), Leibniz (1678), Huygens (1690, preface), Whewell (1847, vol. 2, p. 64f.), Peirce (1883), Duhem (1914, ch. II, §5), Popper (1963, p. 241f.), and Kuhn (1962, p. 154f.). But why should the predictivist thesis be true? In (Maher 1988, 1990a) I show that there is a Bayesian explanation. I do not know of any cogent non-Bayesian explanation of the predictivist thesis.[2]

For Bayesian explanations of some other judgments that scientists make about confirmation, see (Howson and Urbach 1989, ch. 4).

4.2 NORMATIVITY

I have said Bayesian theory can explain judgments that scientists make about when evidence confirms theories. But in what sense does it explain these judgments? One possibility is that the Bayesian explanation represents the reasoning by which scientists arrive at their judgments about confirmation. But this view faces serious difficulties. For one thing, Bayesian explanations of judgments about confirmation often involve moderately lengthy mathematical derivations – my own explanation of the predictivist thesis takes several pages of mathematics. It

---

[2]For criticism of some non-Bayesian attempts at explaining the predictivist thesis, see Campbell and Vinci (1983). Campbell and Vinci (1982, 1983) attempt their own Bayesian explanations, which I criticized in (Maher 1988, p. 283). Howson and Franklin (1991) criticize my defense of the predictivist thesis, but their criticism rests on a failure to distinguish between a hypothesis and the method by which the hypothesis was generated (Maher in press).

is not plausible to think that scientists are doing such nontrivial mathematics when they make judgments of confirmation. Second, psychologists have amassed evidence that people generally do not reason using probability theory, but rather use heuristics that violate probability theory in systematic ways (Kahneman, Slovic, and Tversky 1982).

So I think it is a mistake to view Bayesian confirmation theory as a psychological theory of scientists. I suggest we should view it rather as a normative theory. A Bayesian "explanation" of some judgment scientists make about confirmation purports to show that, and why, this judgment is *rational.* The reason why the judgment is rational may or may not be the reason why the judgment was actually made. Bayesian confirmation theory is concerned with the former, while the latter is the province of psychology.

Why does it matter whether Bayesian confirmation theory can explain the judgments about confirmation that scientists make? Following Rawls (1971, pp. 19–21, 46–7) and Goodman (1973, pp. 62–6), it is now widely allowed that a normative theory should be judged by both (a) the plausibility of its basic assumptions and (b) the degree to which it fits pretheoretic judgments about particular cases. I have argued in Chapters 1 to 3 that the subjective probability theory on which Bayesian confirmation theory rests is plausible; and in Chapter 5 I will show that under fairly weak conditions, the principle of conditionalization follows from this. Thus Bayesian confirmation theory is in a satisfactory state with respect to criterion (a). The point of explaining scientists' intuitive judgments about confirmation is to show that Bayesian confirmation theory is also successful with respect to criterion (b) – it fits pretheoretic judgments about particular cases.

But, some philosophers still ask, how can showing that a theory fits actual judgments show that it is correct as a normative theory? Isn't this a futile attempt to argue from the descriptive to the normative, from what is to what ought to be? Not at all, because the explanation of a judgment of confirmation is only reason to think the theory normatively correct if we are antecedently confident that the judgment itself is correct. So the argument is from particular normative judgments to general ones, not from the descriptive to the normative.

All of this has a more familiar parallel in logic. A system of logic can be used to explain why certain accepted arguments are valid. This explanation cannot generally be regarded as a description of the reasoning of people who make or endorse those arguments. All we can safely say is that the logical explanation of the validity of an argument explains why the judgment of validity is correct. Furthermore, if the question of the correctness of the logical system arises, a relevant factor in answering that question is the extent to which the system fits pretheoretic judgments about the validity of particular arguments.

I have been stressing that Bayesian confirmation theory should be interpreted as normative, not as a descriptive theory of how scientists actually reason. But this is not to say that there is no connection at all between the demonstrations of Bayesian confirmation theory and actual scientific reasoning. Indeed, the remarkable degree of agreement between considered scientific judgments of confirmation, and Bayesian confirmation theory, makes it overwhelmingly likely that there is some connection. But the degree of fit is surely rather crude, in general.

4.3   BETTING ON THEORIES

Bayesian confirmation theory assumes that scientists have subjective probabilities for scientific theories. In Chapter 1, I advocated a preference interpretation of subjective probability (and utility), according to which you have subjective probability function p if your preferences maximize expected utility relative to p (and some utility function). I intend this interpretation to be quite general, and so to apply in particular to subjective probabilities for scientific theories.

It has sometimes been claimed that the preference interpretation of subjective probability cannot apply to probabilities of scientific theories.[3] Those who make this claim seem to have in mind an argument like the following.

1. The preference interpretation of probability defines the probability of A in terms of preferences regarding bets on A.

---

[3]For example, by Shimony (1970, p. 93), Dorling (1972, p. 184), and Thagard (1988, p. 98).

2. A bet on A is possible only if we can definitively establish whether or not A is true.
3. We cannot definitively establish whether or not scientific theories are true.
Therefore:
4. Probabilities for scientific theories cannot be defined by the preference interpretation.

But I will show that this argument has two flaws: It is invalid, and one of its premises is false.

To see that the argument is invalid, note that what follows from (2) and (3) is merely that we cannot actually *implement* bets on scientific theories; it does not follow that we cannot have *preferences* regarding these bets. But it is the impossibility of having preferences regarding bets on scientific theories that is needed to derive the conclusion (4). Thus the argument tacitly assumes

5. Preferences are defined only for possible bets.

But (5) is surely false; we often have preferences between options we could not actually be presented with.

Consider an example: Thagard (1988, p. 98) writes that he "is unable to attach much sense to the question of how to bet, for example, on the theory of evolution." Now there is certainly a problem here, in that "the theory of evolution" is not a well-defined entity. But for present purposes, let's identify this "theory" with the hypothesis that all species on earth evolved from a small number of primitive species. I asked several biologists to imagine that we will definitively determine the truth value of this thesis. (For vividness, I suggested they might imagine that some reliable extraterrestrials have been observing the earth since its early days, and will inform us of what they saw.) I asked these biologists to suppose also that they have their choice of (a) $1,000 if the evolutionary hypothesis is true or (b) $1,000 if it is false. The biologists had no trouble making sense of the imagined situation and were emphatic that they would choose (a). From this we can infer that they *prefer* bet (a) to bet (b), and this is consistent with (a) and (b) not being bets one could actually make.[4]

---

[4]If we assume that the biologists' preference for (a) over (b) maximizes their expected utility, then we can infer that these biologists give evolution a probability greater than 1/2.

I am not claiming that we have preferences regarding all conceivable bets. There may be some pairs of bets for which we neither prefer one to the other, nor are indifferent between them (without violating any requirements of rationality). In such a case, we have indeterminate subjective probabilities and utilities. What I am claiming is that we *can* have preferences regarding conceivable bets that cannot in fact be implemented; and I think most of us do have preferences about many such bets.

I have shown that the argument for (4) is invalid and that this invalidity cannot be removed without adding a false premise. I will now show that the original argument already contains a false premise. This is the premise (2), which asserts that a bet on A is possible only if we can definitively establish whether or not A is true. To see that this is false, note that a bet on A is an arrangement in which some desirable consequence occurs just in case A is true. If we cannot definitively establish whether or not A is true, then we cannot definitively establish whether or not we have won the bet; but it does not follow that we cannot *make* the bet.

This point can be illustrated by the arrangements David Hume made for publishing his *Dialogues Concerning Natural Religion*. Hume had been persuaded by his friends that publication of the *Dialogues* was likely to outrage the religious establishment, and that this could have unfortunate consequences for him. But Hume wanted the *Dialogues* published because he thought they might have a beneficial influence, and because he wanted to add to his fame. So he withheld the *Dialogues* from publication during his lifetime but made elaborate provisions to ensure that they would be published after his death. In making these provisions, Hume was betting that these provisions would result in publication of the *Dialogues*; but Hume knew he would never be able to establish definitively whether or not he had won this bet.[5]

---

[5]Someone will think: "Hume made these arrangements because the thought that the *Dialogues* would be published pleased him – and he could verify that he had this thought." But what Hume wanted to achieve was not that he *think* the *Dialogues* would be published; he sought rather to achieve that the *Dialogues* be published. Given a choice between (i) falsely believing that the *Dialogues* would be published, and (ii) falsely believing that they would not be published, I am sure Hume would have chosen (ii).

In Chapter 6 I will introduce the notion of *accepting* a proposition. Suppose, as I think is not too far from the truth, that scientists typically would like the theories they accept to be true.[6] Thus acceptance of theory A is an arrangement in which scientists get the consequence "Accepting A when it is true" if A is true, and "Accepting A when it is false" if A is false. They regard the former consequence as more desirable than the latter consequence, and so this is a bet on A – even though nobody can definitively establish whether A is true. Since scientists can (and do) accept scientific theories, they can (and do) make bets on propositions whose truth value cannot be conclusively ascertained. Thus (2) is false.

In Chapter 8 I will show that preferences over the cognitive options of accepting various theories, together with conducting various experiments, are all that is needed in order to apply the preference interpretation of probability to scientific theories.

So the preference interpretation of subjective probability is applicable to the subjective probabilities of scientific theories.

4.4 SUBJECTIVITY

Probably the most common objection raised against Bayesian philosophy of science is that it is "too subjective." A vivid example is the following passage by Wesley Salmon. Salmon is discussing the view that satisfaction of the axioms of probability is necessary and sufficient for rationality, a view he associates with Ramsey and de Finetti (Salmon 1967, n. 102). He objects that on this view:

You cannot be convicted of irrationality as long as you are willing to make the appropriate adjustments elsewhere in your system of beliefs. You can believe to degree 0.99 that the sun will *not* rise tomorrow. You can believe with equal conviction that hens will lay billiard balls. You can maintain with virtual certainty that a coin that has consistently come up heads three quarters of the time in a hundred million trials is heavily biased for *tails!* There is no end to the plain absurdities that qualify as rational. It is not that the theory demands the

[6]It would make no difference if we said they want theories they accept to be approximately true, or to be empirically adequate. For we cannot definitively establish whether a theory has these properties, just as we cannot definitively establish whether it is true.

acceptance of such foolishness, but it does tolerate it. (Salmon 1967, p. 81)

This objection can be read in two ways. On one reading, what it claims is that *you*, a person who thinks that the sun will rise tomorrow, and so on, could arbitrarily switch to thinking that the sun will not rise, and so on, without being deemed irrational by this theory. That is a relevant objection to the view that satisfaction of the axioms of probability is sufficient for rationality. But it is not a relevant objection to the Bayesian confirmation theory described in this chapter. The reason is that these arbitrary shifts would violate the principle of conditionalization. Since Ramsey and de Finetti also endorsed conditionalization, it is likewise not a relevant objection to their views.[7]

On the other possible reading of Salmon's objection, what it claims is that a being who had these bizarre probabilities (perhaps by being born with them) would not necessarily be deemed irrational by this theory. In this case, it is not so obvious to me that we are dealing with a case of irrationality. Still, I don't wish to deny the point, that a person who satisfies the formal requirements of Bayesian theory may nevertheless be irrational. I take Bayesian theory to be specifying necessary conditions of rationality, not sufficient ones (compare Section 1.9). So again, my response to Salmon's objection would be that the position he is criticizing is not the one I am defending. I think it is also not the position Ramsey defended. Ramsey (1926, sec. 5) says that logic, which he conceives as the rules of rational thought, contains not only a "logic of consistency" (including probability theory), but also a "logic of truth" (which specifies what inference patterns are reliable).

Perhaps it will now be objected that Bayesian theory is incomplete, because it does not give a sufficient condition of rationality. But I am pessimistic about the prospects for a formal theory that would give a sufficient condition of rationality. Of course, those who think otherwise are welcome to pursue that research program. In any case, it is enough for my purposes here if Bayesian theory gives a necessary condition of rationality.

---

[7]Kyburg has claimed that Bayesians have no good rationale for conditionalization, and that it is inconsistent with the basic approach of Bayesian theory. These claims will be shown to also be mistaken, in Section 5.2.

4.5 EMPIRICISM

According to Bayesian confirmation theory, a rational person's subjective probability function is derived from an earlier subjective probability function, updated by what has been learned in the interim. But this process obviously cannot go back forever. Idealizing somewhat, we ultimately reach a subjective probability function that the person was born with.

The probabilities that a person has today will be to some extent dependent on what that initial probability function was. There are theorems to show that, under certain conditions, and given enough evidence, the effect of the initial probability function will be negligible. While these theorems are significant, the conditions that they require will not always be satisfied.

The upshot is that according to Bayesian confirmation theory, a person's probabilities today are not solely based on experience, but are also partly a function of an initial probability distribution that is unshaped by the person's experience. This situation is unacceptable to strict empiricists, who want all opinion about contingent facts to be based on experience. Locke and Hume were empiricists of this ilk; more recently, van Fraassen has advocated a similar position.[8]

I would answer this objection by simply observing that the only way to conform to the strict empiricist standard is to have no opinions at all about contingent facts, except perhaps those that have been directly observed. For any method of reasoning beyond direct experience cannot itself be based on experience and, in fact, performs the same role as the Bayesian's prior probability function. Since empiricists are plainly unwilling to accept the skepticism their position entails, their objections cannot be taken seriously. (An equally serious objection would be that Bayesians do not base their probabilities solely on the natural light of reason.)

It is sometimes thought that any opinion about contingent fact, if not based on experience, must be "synthetic a priori." Thus Kyburg (1968) argues that Bayesian prior probability

---

[8] "Nothing except experience may be treated as a source of information (hence, prior opinion cannot have such a status, so radical breaks with prior opinion – 'throwing away priors,' as born-again Bayesians call it – can be rational procedure)" (van Fraassen 1985, p. 250).

functions commit one to a priori synthetic knowledge. This is at best misleading. A priori knowledge, as traditionally understood, is not only not based on experience, but is also not revisable by experience. A subjective probability function, even if not based on experience, *is* revisable by experience. Indeed, Bayesian confirmation theory is precisely a theory about how subjective probabilities should be revised in the light of experience. In this sense, subjective probabilities are not a priori.

So while Bayesian confirmation theory is not strictly empiricist, I think it captures as much of empiricism as is reasonable.

4.6 THE DUTCH BOOK ARGUMENT FOR PROBABILITY

I began this chapter by noting the fundamental assumption of Bayesian philosophy of science, namely that scientists have subjective probabilities for scientific theories. Since the theory is normative, a more careful formulation of this fundamental assumption would be that rationality requires scientists to have (not necessarily precise) subjective probabilities for scientific theories. On the preference interpretation of subjective probability that I favor, this fundamental assumption is to be defended by producing a representation theorem and arguing that most scientists are committed to the basic conditions on preference assumed by this theorem.

This way of defending the fundamental assumption of Bayesian philosophy of science is not the usual one adopted by Bayesian philosophers of science. Typically, Bayesian philosophers of science appeal instead to the *Dutch book argument* (Horwich 1982, Howson and Urbach 1989). This argument does have the advantage of being much shorter than any representation theorem. So I need to explain why I invoke a representation theorem, not the Dutch book argument. My reason is that the Dutch book argument is fallacious, as I will now show.

My task here is complicated by the fact that "the Dutch book argument" has been propounded by different authors in different ways; so it is not really one argument but rather a family of resembling arguments. The approach I will adopt is to consider first a definite and simple form of the argument and identify the fallacy as it occurs in this form. I will subsequently

94

show that various refinements of this argument either (a) do not avoid the fallacy, or (b) replace it with another fallacy, or (c) really abandon the Dutch book approach in favor of the representation theorem approach.

4.6.1 The simple argument

First, some gambling terminology: If you pay r for the right to receive s if A is true, you are said to have made a bet on A with a *betting quotient* of r/s, and *stake* s. Here r may be positive, zero, or negative; and s may be positive or negative.

Dutch book arguments normally assume that for any proposition A, there is a number $p(A)$ such that you are willing to accept any bet with betting quotient $p(A)$. As the notation indicates, $p(A)$ is thought of as your subjective probability for A; and the task of the Dutch book argument is to show that it deserves to be called a probability, by showing that rationality requires p to satisfy the axioms of probability.

Dutch book arguments purport to do this by showing that if p does not satisfy the axioms of probability, then you will be willing to accept bets that necessarily give you a loss. A set of bets with this property is called a *Dutch book;* hence the name *Dutch book argument.*[9] Since the argument has been given in full in many places, I will here merely give an example of how it is supposed to work. For this example, let H denote that a coin will land heads on the next toss, and suppose that for you $p(H) = p(\bar{H}) = .6$, which violates the axioms of probability. The Dutch book argument, applied to this case, goes as follows: Since $p(H) = .6$, you are willing to pay 60 cents for a bet that pays you $1 if H. The same holds for $\bar{H}$. But these two bets together will result in you losing 20 cents no matter how the coin lands; they constitute a Dutch book. The Dutch book argument assumes that you do not want to give away money to a bookie, and concludes that your willingness to accept these two bets shows you are irrational.

---

[9]The term 'Dutch book' was introduced to the literature by Lehman (1955), who stated that the term was used by bookmakers.

4.6.2 The fallacy

Suppose that, after a night on the town, you want to catch the bus home. Alas, you find that you have only 60 cents in your pocket, and the bus costs $1. A bookie, learning of your plight, offers you the following deal: If you give him your 60 cents, he will toss a coin; and if the coin lands heads, he will give you $1; otherwise, you have lost your 60 cents. If you accept the bookie's proposal, you stand a 50-50 chance of being able to take the bus home, while rejecting it means you will certainly have to walk. Under these conditions, you may well feel that the bookie's offer is acceptable; let us suppose you do. Presumably the offer would have been equally acceptable if you were betting on tails rather than heads; there is no reason, we can suppose, to favor one side over the other.

As subjective probability was defined for the simple Dutch book argument, your probability for heads in the above scenario is at least .6, and so is your probability for tails; thus your probabilities violate the probability calculus. The Dutch book argument claims to deduce from this that you are irrational. And yet, given the predicament you are in, your preferences seem perfectly reasonable.

Looking back over the simple Dutch book argument I gave, we can see what has gone wrong. From the fact that you are willing to accept each of two bets that together would give a sure loss, that argument infers that you are willing to give away money to a bookie. This assumes that if you are willing to accept each of the bets, you must be willing to accept both of them. But that assumption is surely false in the present case. In being willing to bet at less than even money on either heads or tails, you are merely being sensible; but you would certainly have taken leave of your senses if you were willing to accept both bets together. Accepting both bets, like accepting neither, means you will have to walk home.

This, then, is the fallacy in the Dutch book argument: It assumes that bets that are severally acceptable must also be jointly acceptable; and as our example shows, this is not so.

Now as I mentioned before, there are many versions of the Dutch book argument. Fans of one or another of these versions are sure to protest that the problem just identified is only a problem for the simple form of the argument I illustrated, and is not shared by their favorite version of the argument. But I think that response would be mistaken and that the elementary fallacy just identified actually goes to the core of the Dutch book approach. In the following subsections, I will substantiate this claim by considering the main refinements of the Dutch book argument known to me.

4.6.3 Introduction of utilities

One refinement of the simple Dutch book argument is to define betting quotients in terms of utilities rather than monetary amounts (Shimony 1955, Horwich 1982). In this version of the argument, if you pay r for the right to receive s if A is true, and you have utility function u, then the betting quotient is said to be $u(\$r)/u(\$s)$. This will in general be different from the value r/s that was taken to be the betting quotient in the simple argument. Apart from that difference, everything proceeds as before: It is assumed that for any proposition A, there is a number $p(A)$ such that you are willing to accept any bet with betting quotient $p(A)$ – betting quotients now being defined in the way just stated. And it is argued that if the function p so defined does not satisfy the axioms of probability, then you are willing to accept a set of bets that together produce a sure loss.

Note that this approach assumes that a rational person does have a utility function. A skeptic could well question this assumption. In fact, prior to the representation theorem of von Neumann and Morgenstern (1947), it was standardly denied that utilities had any more than ordinal significance. This kind of skepticism can be answered, as von Neumann and Morgenstern answered it, by giving a representation theorem. But if Dutch book arguments appeal to a representation theorem, they have lost their putative advantage of simplicity over representation theorems.

Even if the assumption of utilities is granted, a more serious problem remains, namely: The Dutch book argument with utilities is still fallacious, in exactly the same way as the simple argument. For even with the introduction of utilities, the argument still assumes that bets that are severally acceptable are

97

jointly acceptable, and we have seen that this is not a requirement of rationality. To see that this is so, return to the example of Section 4.6.2, and suppose your utility function u is such that

$$u(\$0) = 0; \quad u(\$0.40) = .4; \quad u(\$0.60) = .6; \quad u(\$1.00) = 2.$$

Since you currently have $0.60, the expected utility of not accepting the bookie's deal is .6. Assuming your subjective probability for heads is $1/2$, the expected utility of accepting the bookie's offer to bet on the coin landing heads is 1. (You have $1 if the coin lands heads, and nothing otherwise.) Thus you are willing to accept this bet. And the same is true if the bet is on tails rather than heads. But to accept both bets would leave you with only $0.40, and since the utility of that (.4) is less than the status quo, you are not willing to accept both bets. From this perspective, the reason why bets that are severally acceptable need not be jointly acceptable is that utility need not be a linear function of money. More generally, utilities need not be additive: The utility of two things together need not be equal to the sum of the utilities of each separately.

One might try to repair the argument by requiring that the bets considered be for some commodity in which utilities are additive. But as Schick (1986) argues, one would then need a reason to think that such a commodity exists, and no such reason is in sight.[10]

Besides not fixing the fallacy in the simple argument, the introduction of utilities creates a new fallacy that was not present in the simple argument. To see this, suppose your probabilities (defined with betting quotients in terms of utilities) are such that $p(H) = p(\bar{H}) = .6$. The argument that this violation of the probability calculus is irrational is now that you are willing to pay .6 utiles (units of utility) for a bet that pays 1 utile if H is true; and you would pay the same for a bet that pays 1 utile if H is false; and both bets together produce a sure loss. We have already seen that you need not be willing to accept both bets together. The point I wish to make now is that even if you

---

[10]If a person's probabilities satisfy the axioms of probability, then it can be shown that (expected) utilities are additive over lottery tickets. But the Dutch book argument is trying to show that probabilities should satisfy the axioms, so it would beg the question to rely on this fact here.

did accept both bets together, *you need not suffer a loss.* For example, suppose your utilities are such that

$$u(\$1) = 1; \quad u(-\$0.25) = -.6.$$

Then to pay .6 utiles for a bet that pays 1 utile if H (or $\bar{H}$) is to pay 25 cents for a bet that yields \$1 if H (or $\bar{H}$). Thus to accept both bets is to make a sure profit (of 50 cents), not a sure loss.

4.6.4 Posting betting quotients

In de Finetti's formulation of the Dutch book argument, one considers a (hopefully fictitious) situation in which you are required to state some number $p(A)$ for each event A, and must then accept whatever bets anyone cares to make with you, provided the betting quotients agree with your stated probabilities.[11] It is then argued that if the betting quotients you post do not satisfy the axioms of probability, bets can be made that you are obliged to accept and that together will result in you suffering a sure loss.

This version of the Dutch book argument eliminates by fiat the possibility that you will not be willing to accept jointly bets that are severally acceptable. You are now *required* to accept all bets at your stated betting quotients. And if the betting quotient is defined in terms of money rather than utility (as it was in de Finetti's presentation), the difficulties raised by the introduction of utilities are also avoided.

On the other hand, de Finetti's version of the Dutch book argument introduces new difficulties. First, we can question the assumption that it would be irrational for you to set betting quotients such that someone could make a Dutch book against you. This need not be so. Consider again the scenario in which you only have 60 cents, and need \$1 to ride the bus home. You might know the following things:

- If you post betting quotients that satisfy the axioms of probability, then the bookie will make no bet with you (and so you will have to walk home).

[11]See (de Finetti 1931, pp. 304, 308, and 1937, p. 102). The relevant passages from the former article are translated in (Heilig 1978, p. 331).

- If you post betting quotients of .6 for both heads and tails, the bookie will place a bet on one or the other, but not both.

Since you prefer a bet at the betting quotient of .6 to no bet, it is then rational for you to post betting quotients of .6 on heads and tails. Hence it can be rational to post betting quotients that leave you open to a Dutch book.

The preceding difficulty might be avoided by stipulation. De Finetti has already stipulated that the situation being considered is one in which you are obliged to post betting quotients and accept all bets that anyone may care to make at these betting quotients. Why not add the further stipulation that there is a bookie who will make a Dutch book against you if that is possible? With this additional stipulation, we have that you will indeed suffer a sure loss if your posted betting quotients do not satisfy the axioms of probability.

Suppose we make this stipulation. Still it does not follow that you would be irrational to post betting quotients that violate the axioms of probability. For example, suppose you know that if you post betting quotients of .6 on a coin landing heads, and .6 on it landing tails, then the bookie will make bets with you such that you lose $1 for sure, and no other bets will be made. Suppose you also know that if you post betting quotients of .5 on heads and .5 on tails, the bookie will bet you $100 that the coin will land heads. If you are risk averse, you might well prefer to accept the sure loss of $1 than to gamble on losing $100; hence it would be rational for you to post betting quotients that violate the probability calculus.

Of course, this gap in the argument can be closed by a further stipulation. We can suppose you are sure, not only that the bookie will make you suffer a sure loss if you post betting quotients violating the probability calculus, but also that the bookie will bankrupt you. I am willing to concede, at least for the sake of argument, that with all these stipulations assumed, it would indeed be irrational for you to post betting quotients that violate the axioms of probability. (If my concession is premature, some further stipulations can always be introduced to close the gaps.) But, I shall argue, this superficially successful defense of de Finetti's argument achieves a merely Pyrrhic victory.

100

Imagine someone arguing that subjective probabilities must satisfy the axioms of probability because, if you are required to assign numbers to propositions, and are told you will be shot if they do not satisfy the axioms of probability, then you would be irrational to assign numbers that violate the axioms. Clearly this argument has no force at all. (An exactly parallel argument could be used to "show" that subjective probabilities must *violate* the axioms of probability: Just stipulate that if the numbers you assign satisfy the axioms, you will be shot.) The flaw in the argument is that it assumes the numbers a person assigns to propositions in some arbitrarily specified situation can be identified with the person's subjective probabilities; and plainly that is not so.

But now let us ask whether de Finetti's argument does not have the same flaw. When it is stipulated that

- You must post betting quotients and accept all bets anyone wants to make at your posted betting quotients; and
- there is a bookie who will bankrupt you if you post betting quotients that do not satisfy the axioms of probability

is there any good reason to think the betting quotients you specify represent your subjective probabilities? To answer this question, we need to think about what makes something your subjective probability. In Chapter 1, I suggested that for a function p to be your subjective probability function, it must be the case that the attribution of p and u to you provides a sufficiently good interpretation of your preferences. And a central desideratum for an interpretation is that it represent your preferences as by and large maximizing expected utility.

Now the numbers that you would be rational to nominate, when all the stipulations de Finetti must make are assumed, need not be your subjective probabilities, in this sense. For example, suppose your true subjective probabilities are .6 for H and $\bar{H}$. Given the stipulations de Finetti must make, you know that if you post betting quotients of .6 for both these events, you will be bankrupted. You also figure that if you post betting quotients of .5 for both events, you stand at most a .6 chance of being bankrupted. (You are bankrupted iff (a) the bookie bets an amount equal to your total wealth on H or $\bar{H}$, but not

101

both; and (b) the bookie wins the bet. You put the probability of (b) at .6.) Hence you see that you would be irrational to post betting quotients equal to your true subjective probabilities.

Thus de Finetti's argument is fallacious, in the same way as the argument that appeals to what you would do when a gun is put to your head. At best, the argument shows that, under some specified conditions, it would be rational for you to assign numbers to propositions in a way that satisfies the axioms of probability. Since these numbers need not be your subjective probabilities, it does not follow that rationality requires you to have subjective probabilities satisfying the axioms of probability.

4.6.5 Mathematical aggregation

Skyrms (1987a) has recently claimed that the Dutch book argument can be reformulated in a way that makes it cogent. He refers to the result of making both bets b_1 and b_2 as the *physical aggregate* of b_1 and b_1. This is the kind of combination of bets that has been considered in the versions of the Dutch book argument discussed so far. Skyrms proposes to consider instead what he calls *mathematical* aggregates of bets. The mathematical aggregate of b_1 and b_2 is an arrangement that, in each state, gives a consequence whose utility is equal to the sum of the utilities of the consequences b_1 and b_2 give in that state.

In the first two versions of the Dutch book argument that I considered, it was erroneously assumed that bets that are severally acceptable must be jointly acceptable. In the terminology just introduced, we can put the point by saying that the physical aggregate of a set of acceptable bets need not itself be an acceptable bet.

From the perspective of expected utility theory, the reason why severally acceptable bets need not be jointly acceptable is that the expected utility of the physical aggregate need not be the sum of the expected utilities of the bets involved. On the other hand, the expected utility of a mathematical aggregate is necessarily the sum of the expected utilities of the aggregated bets. Hence if each of a set of bets is acceptable (i.e., each has higher expected utility than the status quo), then the mathematical

102

aggregate of those bets should also be acceptable. So if we re-place physical by mathematical aggregates, a false assumption of Dutch book arguments is replaced by a true one.

However, this also means that the mathematical aggregate of a set of separately acceptable bets must have greater expected utility than the status quo. Thus the mathematical aggregate of a set of separately acceptable bets can never give a sure loss. Hence if physical aggregation is replaced by mathematical aggregation, a Dutch book argument cannot possibly show that a person who violates the probability calculus is willing to accept a sure loss. Nor does Skyrms claim that it could.

What, then, is Skyrms's version of the Dutch book argument? Skyrms makes the following assumptions. (Here $f\#g$ denotes the mathematical aggregate of f and g.)

(1) If b gives the same utility u in every state, then the expected utility of b, $EU(b)$, equals u.

(2) If $EU(b_1) = EU(b'_1)$ and $EU(b_2) = EU(b'_2)$, then $EU(b_1\#b_2) = EU(b'_1\#b'_2)$.

Skyrms shows that given these assumptions, probabilities must satisfy the additive law of probability (i.e., the probability of a disjunction of incompatible propositions must be equal to the sum of the probabilities of the disjuncts).

Skyrms anticipates objections to assumption (2); and he responds that this can be derived from more restrictive principles. He also begins to sketch a representation theorem, à la Ramsey, which would enable the additivity of probability to be derived without assuming (2). He says that "If we pursue this line we, like Ramsey, will have followed the dutch book theorems to deeper and deeper levels until it leads to the representation theorem."

However, what Skyrms here presents as a *defense* of Dutch book arguments, I interpret as an *abandonment* of such arguments, in favor of the representation theorem approach. The point is partly semantic. I think it is most useful to use the term 'Dutch book argument' to refer only to arguments that involve a Dutch book, that is, a set of bets that together produce a sure loss. Skyrms's version of the Dutch book argument involves no Dutch book, so is not really a Dutch book argument at all. Similarly, no Dutch book figures in representation

103

theorems, so representation theorems are not a species of Dutch book argument.

Apart from the semantic issue of what counts as a Dutch book argument, I think Skyrms and I are in complete agreement. He seems to allow that it cannot cogently be argued that violation of the probability calculus makes you susceptible to a Dutch book. And he maintains, as I do, that the cogent way to argue that rational scientists have subjective probabilities, satisfying the axioms of probability, is to use a representation theorem.

5

Diachronic rationality

We have seen that Bayesian confirmation theory rests on two assumptions: (1) That rational scientists have probabilities for scientific hypotheses, and (2) the principle of conditionalization. The latter is a *diachronic* principle of rationality, because it concerns how probabilities at one time should be related to probabilities at a later time.

Chapters 1–4 gave an extended argument in support of (1). This chapter will examine what can be said on behalf of (2). I will reject the common Bayesian view that conditionalization is a universal requirement of rationality, but argue that nevertheless it should hold in normal scientific contexts.

I begin by discussing a putative principle of rationality known as Reflection. A correct understanding of the status of this principle will be the key to my account of the status of conditionalization.

5.1 REFLECTION

Suppose you currently have a (personal) probability function p, and let R_q denote that at some future time $t + x$ you will have probability function q. Goldstein (1983) and van Fraassen (1984) have claimed that the following identity is a requirement of rationality:[1]

$$p(\cdot|R_q) = q(\cdot).$$

Following van Fraassen (1984), I will refer to this identity as *Reflection.*

As an example of what Reflection requires, suppose you are sure that you cannot drive safely after having ten drinks. Suppose further that you are sure that after ten drinks, you would

---

[1] Goldstein actually defends a stronger condition; but the argument for his stronger condition is the same as for the weaker one stated here.

be sure (wrongly, as you now think) that you could drive safely. Then you violate Reflection. For if p is your current probability function, q the one you would have after ten drinks, and D the proposition that you can drive safely after having ten drinks, we have

$$p(D|R_q) \approx 0 < 1 \approx q(D).$$

Reflection requires $p(D|R_q) = q(D) \approx 1$. Thus you should now be sure that you would not be in error, if in the future you become sure that you can drive safely after having ten drinks.

5.1.1 The Dutch book argument

Why should we think Reflection is a requirement of rationality? According to Goldstein and van Fraassen, this conclusion is established by a diachronic Dutch book argument. A diachronic Dutch book argument differs from a regular Dutch book argument in that the bets are not all offered at the same time. But like a regular Dutch book argument, it purports to show that anyone who violates the condition is willing to accept bets that together produce a sure loss, and hence is irrational.

Since the diachronic Dutch book argument for Reflection has been stated in full generality elsewhere (Goldstein 1983, van Fraassen 1984, Skyrms 1987b), I will here merely illustrate how it works. Suppose, then, that you violate Reflection with respect to drinking and driving, in the way indicated above. For ease of computation, I will assume that $p(D) = 0$ and $q(D) = 1$. (Using less extreme values would not change the overall conclusion.) Let us further assume that your probability that you will have ten drinks tonight is $1/2$. The Dutch bookie tries to make a sure profit from you by first offering a bet b_1 whose payoff in units of utility is[2]

$$-2 \text{ if } DR_q; \qquad 2 \text{ if } \bar{D}R_q; \qquad -1 \text{ if } \bar{R}_q.$$

For you at this time, $p(DR_q) = 0$, and $p(\bar{D}R_q) = p(\bar{R}_q) = 1/2$. Thus the expected utility of b_1 is $1/2$. We are taking the utility of the status quo to be 0, and so the bookie figures that you

[2]Conjunction is represented by concatenation, and negation by overbars. For example, $\bar{D}R_q$ is the proposition that D is false and R_q is true.

106

will accept this bet. If you accept the bet and do not get drunk (R_q is false), you lose one unit of utility. If you accept and do get drunk (R_q is true), the bookie offers you b_2, whose payoff in units of utility is

$$1 \text{ if } D; \; -3 \text{ if } \bar{D}.$$

Since you are now certain D is true, accepting b_2 increases your expected utility, and so the bookie figures you will accept it. But now, if D is true, you gain 1 from b_2 but lose 2 from b_1, for an overall loss of 1. And if D is false, you gain 2 from b_1 but lose 3 from b_2, again losing 1 overall. Thus no matter what happens, you lose.[3]

5.1.2 Counterexamples

Despite this argument, there are compelling prima facie counterexamples to Reflection. Indeed, the drinking/driving example is already a prima facie counterexample; it seems that you would be right to now discount any future opinions you might form while intoxicated – contrary to what Reflection requires. But we can make the counterexample even more compelling by supposing you are *sure* that tonight you will have ten drinks. It then follows from Reflection that you should *now* be sure that you can drive safely after having ten drinks.

Proof. $\quad p(D) \;=\; p(D|R_q), \quad \text{since } p(R_q) = 1$
$$ \;=\; q(D), \quad \text{by Reflection}$$
$$ \;=\; 1.$$

This result seems plainly wrong. Nor does it help to say that a rational person should not drink so much; for it may be that the drinking you know you will do tonight will not be voluntary.

A defender of Reflection might try responding to such counterexamples by claiming that the person you would be when drunk is not the same person who is now sober. If you were

---

[3]In presentations of this argument, it is usual to have two bets, where I have the single bet b_1. Those two bets would be a bet on R_q, and a bet on $\bar{D}$ which is called off if R_q is false. By using a single bet instead, I show that the argument does not here require the assumption that bets that are separately acceptable are also jointly acceptable.

absolutely sure of this, then for you $p(R_q) = 0$, since R_q asserts that *you* will come to have probability function q.[4] In that case, $p(\cdot|R_q)$ may be undefined and the counterexample thereby avoided. But this is a desperate move. Nobody I know gives any real credence to the claim that having ten drinks, and as a result thinking he or she can drive safely, would destroy his or her personal identity. They are certainly not sure that this is true.

Alternatively, defenders of Reflection may bite the bullet, and declare that even when it is anticipated that one's probabilities will be influenced by drugs, Reflection should be satisfied. Perhaps nothing is too bizarre for such a die-hard defender of Reflection to accept. But it may be worth pointing out a peculiar implication of the position here being embraced: It entails that rationality requires taking mind-altering drugs, in circumstances where that position seems plainly false. I will now show how that conclusion follows.

It is well known that under suitable conditions, gathering evidence increases the expected utility of subsequent choices, if it has any effect at all. The following conditions are sufficient:[5]

1. The evidence is "cost free"; that is, gathering it does not alter what acts are subsequently available, nor is any penalty incurred merely by gathering the evidence.
2. Reflection is satisfied for the shifts in probability that could result from gathering the evidence. That is to say, if p is your current probability function, then for any probability function q you could come to have as a result of gathering the evidence, $p(\cdot|R_q) = q(\cdot)$.
3. The decision to gather the evidence is not "symptomatic"; that is, it is not probabilistically relevant to states it does not cause.
4. Probabilities satisfy the axioms of probability, and choices maximize expected utility at the time they are made.

[4] I assume that you are sure you cannot come to have q other than by drinking. This is a plausible assumption if (as we can suppose) q also assigns an extremely high probability to the proposition that you have been drinking.

[5] I assume causal decision theory. For a discussion of this theory, and a proof that the stated conditions are indeed sufficient in this theory, see (Maher 1990b). In that work, I referred to Reflection as *Miller's principle*.

108

Now suppose you have the opportunity of taking a drug that will influence your probabilities in some way that is not completely predictable. The drug is cost free (in particular, it has no direct effect on your health or wealth), and the decision to take the drug is not symptomatic. Assume also that rationality requires condition 4 above to be satisfied. If Reflection is a general requirement of rationality, condition 2 should also be satisfied for the drug-induced shifts. Hence all four conditions are satisfied, and it follows that you cannot reduce your expected utility by taking this drug; and you may increase it.

For example, suppose a bookie is willing to bet with you on the outcome of a coin toss. You have the option of betting on heads or tails; you receive $1 if you are right and lose $2 if you are wrong. Currently your probability that the coin will land heads is 1/2, and so you now think the best thing to do is not bet. (I assume that your utility function is roughly linear for such small amounts of money.) But suppose you can take a drug that will make you certain of what the coin toss will be; you do not know in advance whether it will make you sure of heads or tails, and you antecedently think both results equally likely.[6] The drug is cost free, and you satisfy condition 4. Then if Reflection should hold with regard to the drug-induced shifts, you think you can make money by taking the drug. For after you take the drug, you will bet on the outcome you are then certain will result; and if you satisfy Reflection, you are now certain that bet will be successful. By contrast, if you do not take the drug, you do not expect to make a profit betting on this coin toss. Thus the principle of maximizing expected utility requires you to take the drug.

But in fact, it is clear that taking the drug need not be rational. You could perfectly rationally think that the bet you would make after taking the drug has only a 50-50 chance of winning, and hence that taking the drug is equivalent to choosing

---

[6]This condition is necessary in order to ensure that your decision to take the drug is not symptomatic. If you thought the drug was likely to make you sure the coin will land heads (say), and if Reflection is satisfied, then the probability of the coin landing heads, given that you take the drug, would also be high. Since taking the drug has no causal influence on the outcome of the toss, and since the unconditional probability of heads is 1/2, taking the drug would then be a symptomatic act.

109

randomly to bet on heads or tails. Since thinking that violates Reflection, we have another reason to deny that Reflection is a requirement of rationality.

5.1.3 The fallacy

We now face a dilemma. On the one hand, we have a diachronic Dutch book argument to show that Reflection is a requirement of rationality. And on the other hand, we have strong reasons for saying that Reflection is *not* a requirement of rationality. There must be a mistake here somewhere.

In (Maher 1992), following Levi (1987), I argued that a sophisticated bettor who looks ahead will not accept the bets offered in the Dutch book argument for Reflection. The thought was that if you look ahead, you will see that accepting b_1 inevitably leads to a sure loss, and hence will refuse to take the first step down the primrose path. This diagnosis assumed that if you do not accept b_1, you will not be offered b_2. However, Skyrms (in press) points out that the bookie could offer b_2 if R_q obtains, regardless of whether b_1 has been accepted. Faced with this strategy, you do best (maximize expected utility) to accept b_1 as well, and thus ensure a sure loss.

So with Skyrms's emendation, the diachronic Dutch book argument does show that if you violate Reflection, you can be made to suffer a sure loss. Yet as Skyrms himself agrees, it is not necessarily rational to conform to Reflection. Thus we have to say that *susceptibility to a sure loss does not prove irrationality*. This conclusion may appear counterintuitive; but that appearance is an illusion, I will now argue.

We say that act a *dominates* act b if, in every state, the consequence of a is better than that of b. It is uncontroversial that it is irrational to choose an act that is dominated by some other available act. Call such an act *dominated*. One might naturally suppose that accepting a sure loss is a dominated act, and thereby irrational.

But consider this case: I have bet that it will not rain today. The deal, let us say, is that I lose \$1 if it rains and win \$1 otherwise. How I came to make this bet does not matter – perhaps it looked attractive to me at the time; perhaps I made

110

	No rain	Rain
Accept 2nd bet	−$0.50	−$0.50
Don't accept	$1	−$1

Figure 5.1: Available options after accepting bet against rain

it under duress. In any case, storm clouds are now gathering and I think I will lose the bet. I would now gladly accept a bet that pays me $0.50 if it rains and in which I pay $1.50 otherwise. If I did accept such a new bet, then together with the one I already have, I would be certain to lose $0.50. So I am willing to accept a sure loss. But I am not thereby irrational. The sure loss of $0.50 is better than a high probability of losing $1. Note also that although I am willing to accept a sure loss, I am *not* willing to accept a dominated option. My options are shown in Figure 5.1. The first option, which gives the sure loss, is not dominated by the only other available act.

So we see that acceptance of a sure loss is not always a dominated act; and when it is not, acceptance of a sure loss can be rational. I suggest that the intuitive irrationality of accepting (or being willing to accept) a sure loss results from the false supposition that acceptance of a sure loss is always a dominated option, combined with the correct principle that it is irrational to accept (or be willing to accept) a dominated option.

Let us apply this to the Dutch book argument for Reflection. In the example of Section 5.1.1, you are now certain that accepting b_2 would result in a loss, and hence you prefer that you not accept it. However, you also know that you will accept it if you get drunk. This indicates that your willingness to accept b_2 when drunk is not something you are now able to reverse (for if you could, you would). Thus you are in effect now stuck with the fact that you will accept b_2 if you get drunk, that is, if R_q is true. Hence you are in effect now saddled with the bet

$$1 \text{ if } DR_q; \quad -3 \text{ if } \bar{D}R_q,$$

111

	DR_q	$\bar{D}R_q$	$\bar{R}_q$
Accept b_1	−1	−1	−1
Reject b_1	1	−3	0

Figure 5.2: Available options for Reflection violator

though it looks unattractive to you now. (This is analogous to the first bet in the rain example.) But you do have a choice about whether or not to accept b_1. Since b_1 pays

$$-2 \text{ if } DR_q; \qquad 2 \text{ if } \bar{D}R_q; \qquad -1 \text{ if } \bar{R}_q$$

your options and their payoffs are as in Figure 5.2. Accepting b_1 ensures that you suffer a sure loss; but it is not a dominated option. In fact, since $p(D) = 0$ and $p(R_q) = 1/2$, accepting b_1 reduced your expected loss from −1.5 to −1. So in this case, as in the rain example, the willingness to accept a sure loss does not involve willingness to accept a dominated option, and does not imply irrationality.

If there is any irrationality in this case, it lies in the potential future acceptance of b_2. But because that future acceptance is outside your present control, it is no reason to say that you are now irrational. Perhaps your future self would be irrational when drunk, but that is not our concern. Reflection is a condition on your present probabilities only, and what we have seen is that you are not irrational to now have the probabilities you do, even though having these probabilities means you are willing to accept a sure loss.

Let us say that a Dutch book *theorem* asserts that violation of some condition leaves one susceptible to a sure loss, while a Dutch book *argument* infers from the theorem that violation of the condition is irrational. In Section 4.6, the condition in question was satisfaction of the axioms of probability, and my claim in effect was that the argument fails because the theorem is false. In the present section, the condition in question has been Reflection, and my claim has been that here the Dutch book theorem is correct but the argument based on it is

112

fallacious. Consequently, this argument provides no reason not to draw the obvious conclusion from the counterexamples in Section 5.1.2: Reflection is not a requirement of rationality. (Christensen [1991] and Talbott [1991] arrive at the same conclusion by different reasoning.)

5.1.4 Integrity

Recognizing the implausibility of saying Reflection is a requirement of rationality, van Fraassen (1984, pp. 250–5) tried to bolster its plausibility with a voluntarist conception of personal probability judgments. He claimed that personal probability judgments express a kind of commitment; and he averred that integrity requires you to stand behind your commitments, including conditional ones. For example, he says your integrity would be undermined if you allowed that were you to promise to marry me, you still might not do it. And by analogy, he concludes that your integrity would be undermined if you said that your probability for A, given that tomorrow you give it probability r, is something other than r.

I agree that a personal probability judgment involves a kind of commitment; to make such a judgment is to accept a constraint on your choices between uncertain prospects. For example, if you judge A to be more probable than B, and if you prefer $1 to nothing, then faced with a choice between

(i) $1 if A, nothing otherwise

and

(ii) nothing if A, $1 otherwise,

you are committed to choosing (i). But of course, you are not thereby committed to making this choice at all times in the future; you can revise your probabilities without violating your commitment. The commitment is to make that choice *now*, if *now* presented with those options. But this being so, a violation of Reflection is not analogous to thinking you might break a marriage vow. To think you might break a marriage vow is to think you might break a commitment. To violate Reflection is to not *now* be committed to acting in accord with a future

113

commitment, on the assumption that you will in the future have that commitment. The difference is that in violating Reflection, you are not thereby conceding that you might ever act in a way that is contrary to your commitments at the time of action. A better analogy for violations of Reflection would be saying that you now think you would be making a foolish choice, if you were to decide to marry me. In this case, as in the case of Reflection, you are not saying you could violate your commitments; you are merely saying you do not now endorse certain commitments, even on the supposition that you were to make them. Saying this does not undermine your status as a person of integrity.

5.1.5 Reflection and learning

In the typical case of taking a mind-altering drug, Reflection is violated, and we also feel that while the drug would shift our probabilities, we would not have *learned* anything in the process. For instance, if a drug will make you certain of the outcome of a coin toss, then under typical conditions the shift produced by the drug does not satisfy Reflection, and one also does not regard taking the drug as a way of *learning* the outcome of the coin toss.

Conversely, in typical cases where Reflection is satisfied, we do feel that the shift in probabilities would involve learning something. For example, suppose Persi is about to toss a coin, and suppose you know that Persi can (and will) toss the coin so that it lands how he wants, and that he will tell you what the outcome will be if you ask. Then asking Persi about the coin toss will, like taking the mind-altering drug, make you certain of the outcome of the toss. But in this case, Reflection will be satisfied, and we can say that by asking Persi you will *learn* how the coin is going to land.

What makes the difference between these cases is not that a drug is involved in one and testimony in the other. This can be seen by varying the examples. Suppose you think Persi really has no idea how the coin will land but has such a golden tongue that if you talked to him you would come to believe him; in this case, a shift caused by talking to Persi will not satisfy Reflection, and you will not think that by talking to him you will learn the

114

outcome of the coin toss (even though you will become sure of some outcome). Conversely, you might think that if you take the drug, a benevolent genie will influence the coin toss so that it agrees with what the drug would make you believe; in this case, the shift in probabilities caused by taking the drug will satisfy Reflection, and you will think that by taking the drug you will learn the outcome of the coin toss.

These considerations lead me to suggest that regarding a potential shift in probability as a learning experience is the same thing as satisfying Reflection in regard to that shift. Symbolically: You regard the shift from p to q as a learning experience just in case $p(\cdot|R_q) = q(\cdot)$.[7]

Shifts that do not satisfy Reflection, though not learning experiences in the sense just defined, may still involve some learning. For example, if q is the probability function you would have after taking the drug that makes you sure of the outcome of the coin toss, you may think that in shifting to q you would learn that you took the drug but not learn the outcome of the coin toss. In general, what you think you would learn in shifting from p to q is represented by the difference between p and $p(\cdot|R_q)$.[8] When Reflection is satisfied, what is learned is represented by the difference between p and q, and we call the whole shift a learning experience.

Learning, so construed, is not limited to cases in which new empirical evidence is acquired. You may have no idea what is the square root of 289, but you may also think that if you pondered it long enough you would come to concentrate your probability on some particular number, and that potential shift may well satisfy Reflection. In this case, you would regard the potential shift as a learning experience, though no new empirical evidence has been acquired. On the other hand, any shift in probability

[7]This proposal was suggested to me by Skyrms (1990a), who assumes that what is thought to be a learning experience will satisfy Reflection. (He calls Reflection "Principle (M)".)

[8]This assumes that q records everything relevant about the shift. Otherwise, it would be possible to shift from p to q in different ways (e.g., by acquiring evidence or taking drugs), some of which would involve more learning than others. This assumption could fail, e.g., if after making the shift, you would forget some relevant information about the shift. Such cases could be dealt with by replacing R_q with a proposition that specifies your probability distribution at every instant between t and $t + x$.

that is thought to be due solely to the influence of evidence is necessarily regarded as a learning experience. Thus satisfaction of Reflection is necessary, but not sufficient, for regarding a shift in probability as due to empirical evidence.

A defender of Reflection might think of responding to the counterexamples by limiting the principle to shifts of a certain kind. But the observations made in this section show that such a response will not help. If Reflection were said to be a requirement of rationality only for shifts caused in a certain way (e.g., by testimony rather than drugs), then there would still be counterexamples to the principle. And if Reflection were said to be a requirement of rationality for shifts that are regarded as learning experiences, or as due to empirical evidence, then the principle would be one that it is impossible to violate, and hence vacuous as a principle of rationality.[9]

5.1.6 Reflection and rationality

Although there is nothing irrational about violating Reflection, it is often irrational to implement those potential shifts that violate Reflection. That is to say, while one can rationally have $p(\cdot|R_q) \neq q(\cdot)$, it will in such cases often be irrational to choose a course of action that might result in acquiring the probability function q. The coin-tossing example of Section 5.1.2 provides an illustration of this. Let H denote that the coin lands heads, and let q be the probability function you would have if you took the drug, and it made you certain of H. Then if you think taking the drug gives you only a random chance of making a successful bet, $p(H|R_q) = .5 < q(H) = 1$, and you violate Reflection; but then you would be irrational to take the drug, since the expected return from doing so is $(1/2)(\$1) - (1/2)(\$2) < 0$.[10]

---

[9] Jeffrey (1988, p. 233) proposed to restrict Reflection to shifts that are "reasonable," without saying what that means. His proposal faces precisely the dilemma I have just outlined. If a "reasonable" shift is defined by its causal origin, Jeffrey's principle is not a requirement of rationality. If a "reasonable" shift is defined to be a learning experience, Jeffrey's principle is vacuous. In the next section, we will see that if a "reasonable" shift is a shift that it would be rational to implement, Jeffrey's principle is again not a requirement of rationality.

[10] Here and in what follows, I assume that the proviso of Section 1.9 holds. That is, I assume your probabilities and utilities are themselves rational, so that rationality requires maximizing expected utility.

This observation can be generalized, and made more precise, as follows. Let d and d' be two acts; for example, d might be the act of taking the drug in the coin-tossing case, and d' the act of not taking the drug. Assume that

(i) Any shift in probability after choosing d' would satisfy Reflection.

In the coin-tossing case, this will presumably be satisfied; if q' is the probability function you would have if you decided not to take the drug, q' will not differ much from p, and in particular $p(H|R_{q'}) = q'(H) = p(H) = .5$.
Assume also that

(ii) d and d' influence expected utility only via their influence on what subsequent choices maximize expected utility.

More fully: Choosing d or d' may have an impact on your probability function, and thereby influence your subsequent choices; but (ii) requires that they not influence expected utility in any other way. So there must not be a reward or penalty attached directly to having any of the probability functions that could result from choosing d or d'; nor can the choice of d or d' alter what subsequent options are available. This condition will also hold in the coin-tossing example if the drug is free and has no deleterious effects on health and otherwise the situation is fairly normal.[11]
Assume further that

(iii) If anything would be learned about the states by choosing d, it would also be learned by choosing d'.

What I mean by (iii) is that the following four conditions are all satisfied. Here Q is the set of all probability functions that you could come to have if you chose d.

(a) You are sure there is a fact about what probability function you would have if you chose d; that is, you give probability 1 to the proposition that for some q, the counterfactual conditional $d \rightarrow R_q$ is true.

---

[11] According to an idea floated in Section 1.8, satisfaction of (ii) ensures that the subjective probabilities that it is rational to have are also justified.

117

(b) For all $q \in Q$ there is a probability function q' such that $p(d' \to R_{q'} | d \to R_q) = 1$.

(c) There is a set S of states of nature that are suitable for calculating the expected utility of the acts that will be available after the choice between d and d' is made. (What this requires is explained in the first paragraph of the proof given in Appendix A.)

(d) For all $q \in Q$, and for q' related to q as in (b), and for all $s \in S$, $p(s|R_q) = p(s|R_{q'})$.

In the coin-tossing example, condition (a) can be assumed to hold: Presumably the drug is deterministic, so that there is a fact about what probability function you would have if you took the drug, though you do not know in advance what that fact is. Condition (b) holds trivially in the coin-tossing example, because not taking the drug would leave you with the same probability function q' regardless of what effect the drug would have. Condition (c) is satisfied by taking $S = \{H, \bar{H}\}$. And it is a trivial exercise to show that (d) holds, since

$$p(H|R_q) = p(H) = 1/2 = p(H|R_{q'}).$$

The coin-tossing example thus satisfies condition (iii). We could say that in this example, you learn nothing about the states whether you choose d or d'.

Also assume that

(iv) d and d' have no causal influence on the states S mentioned in (c).

In the coin-tossing example, neither taking the drug nor refusing it has any causal influence on how the coin lands; and so (iv) is satisfied.

Finally, assume that

(v) d and d' are not evidence for events they have no tendency to cause.

In the coin-tossing example, (iv) and (v) together entail that $p(H|d) = p(H|d') = 1/2$, which is what one would expect to have in this situation.

Theorem. *If conditions (i)–(v) are known to hold, then the expected utility of d' is not less than that of d, and may be greater.*

So it would always be rational to choose d', but it may be irrational to choose d. The proof is given in Appendix A.

The theorem can fail when the stated conditions do not hold. For one example of this, suppose you are convinced there is a superior being who gives eternal bliss to all and only those who are certain that pigs can fly. Suppose also that there is a drug that, if you take it, will make you certain that pigs can fly. If q is the probability function you would have after taking this drug, and F is the proposition that pigs can fly, then $q(F) = 1$. Presumably $p(F|R_q) = p(F) \approx 0$. So the shift resulting from taking this drug violates Reflection. On the other hand, not taking the drug would leave your current probability essentially unchanged. But in view of the reward attached to being certain pigs can fly, it would (or at least, could) be rational to take the drug and thus implement a violation of Reflection.[12] Here the result fails, because condition (ii) does not hold: Taking the drug influences your utility other than via its influence on your subsequent decisions.

To illustrate another way in which the result may fail, suppose you now think there is a 90 percent chance that Persi knows how the coin will land, but that after talking to him you would be certain that what he told you was true. Again letting H denote that the coin lands heads, and letting q_H be the probability function you would have if Persi told you the coin will land heads, we have $p(H|R_{q_H}) = .9$, while $q_H(H) = 1$. Similarly for $q_{\bar{H}}$. Thus talking to Persi implements a shift that violates Reflection. If you do not talk to Persi, you will have probability function q' which, so far as H is concerned, is identical to your current probability function p; so $p(H|R_{q'}) = q'(H) = .5$. Thus not talking to Persi avoids implementing a shift that violates Reflection. Your expected return from talking to Persi is

$$(.9)(\$1) + (.1)(-\$2) = \$0.70.$$

[12]If eternal bliss includes epistemic bliss, taking the drug could even be rational from a purely epistemic point of view. Nevertheless, your certainty that pigs can fly would not be justified (see Section 1.8).

119

Since you will not bet if you do not talk to Persi, the expected return from not talking to him is zero. Hence talking to Persi maximizes your expected monetary return. And assuming your utility function is approximately linear for small amounts of money, it follows that talking to Persi maximizes expected utility. Here the theorem fails because condition (iii) fails. By talking to Persi, you do learn something about how the coin will land; and you learn nothing about this if you do not talk to him. The theorem I stated implies that the expected utility of talking to Persi is no higher than that of learning what you would learn from him, without violating Reflection; but in the problem I have described, the latter option is not available.

I will summarize the foregoing theorem by saying that, other things being equal, implementing a shift that violates Reflection cannot have greater expected utility than implementing a shift that satisfies Reflection. Conditions (i)–(v) specify what is meant here by "other things being equal." This, not the claim that a rational person must satisfy Reflection, gives the true connection between Reflection and rationality.

5.2 CONDITIONALIZATION

In Chapter 4, I noted that Bayesian confirmation theory makes use of a principle of conditionalization. The principle, as I formulated it there, was:

Conditionalization. *If your current probability function is p, and if q is the probability function you would have if you learned E and nothing else, then $q(\cdot)$ should be identical to $p(\cdot|E)$.*

An alternative formulation, couched in terms of evidence rather than learning, will be discussed in Section 5.2.3.

Paul Teller (1973, 1976) reports a Dutch book argument due to David Lewis, which purports to show that conditionalization is a requirement of rationality. The argument is essentially the same as the Dutch book argument for Reflection,[13] and is fallacious for the same reason.

---

[13] But this way of putting the matter reverses the chronological order, since Lewis formulated the argument for conditionalization before the argument for Reflection was advanced.

120

5.2.1 Conditionalization, Reflection, and rationality

In this section, I will argue that conditionalization is not a universal requirement of rationality, and will explain what I take to be its true normative status.

Recall what the conditionalization principle says: If you learn E, and nothing else, then your posterior probability should equal your prior probability conditioned on E. But what does it mean to "learn E, and nothing else"? In Section 5.1.5, I suggested that what you think you would learn in shifting from p to q is represented by the difference between p and $p(\cdot|R_q)$. From this perspective, we can say that you think you would learn E and nothing else, in shifting from p to q, just in case $p(\cdot|R_q) = p(\cdot|E)$.

This is only a subjective account of learning; it gives an interpretation of what it means to *think* E would be learned, not what it means to *really* learn E. But conditionalization is plausible only if your prior probabilities are rational; and then the subjective and objective notions of learning presumably coincide. So we can take the "learning" referred to in the principle of conditionalization to be learning as judged by you. In what follows, I use the term 'learning' in this way.

So if you learned E and nothing else, and if your probabilities shifted from p to q, then $p(\cdot|R_q) = p(\cdot|E)$. If you also satisfy Reflection in regard to this shift, then $p(\cdot|R_q) = q(\cdot)$, and so $q(\cdot) = p(\cdot|E)$, as conditionalization requires. This simple inference shows that Reflection entails conditionalization.

It is also easy to see that if you learn E, and nothing else, and if your probabilities shift in a way that violates Reflection, then your probability distribution is not updated by conditioning on E. For since you learned E, and nothing else, $p(\cdot|R_q) = p(\cdot|E)$; and since Reflection is not satisfied in this shift, $q(\cdot) \neq p(\cdot|R_q)$, whence $q(\cdot) \neq p(\cdot|E)$.

These results together show that conditionalization is equivalent to the following principle: When you learn E and nothing else, do not implement a shift that violates Reflection. But we saw, in Section 5.1.6, that there are cases in which it is rational to implement a shift that violates Reflection. I will now show that some of these cases are ones in which you learn E, and nothing else. This suffices to show that it can be rational to violate conditionalization.

Consider again the situation in which you are sure there is a superior being who will give you eternal bliss, if and only if you are certain that pigs can fly; and there is a drug available that will make you certain of this. Let d be the act of taking the drug, and q the probability function you would have after taking the drug. Then we can plausibly suppose that $p(\cdot|R_q) = p(\cdot|d)$, and hence that in taking the drug you learn d, and nothing else. Consequently, conditionalization requires that your probability function after taking the drug be $p(\cdot|d)$, which it will not be. (With F denoting that pigs can fly, $p(F|d) = p(F) \approx 0$, while $q(F) = 1$.) Hence taking the drug implements a violation of conditionalization. Nevertheless, it is rational to take the drug in this case, and hence to violate conditionalization.

Similarly for the other example of Section 5.1.6. Here you think there is a 90 percent chance that Persi knows how the coin will land, but you know that after talking to him, you would become certain that what he told you was true. We can suppose that in talking to Persi, you think you will learn what he said, and nothing else. Then an analysis just like that given for the preceding example shows that talking to Persi implements a violation of conditionalization. Nevertheless, it is rational to talk to Persi, because (as we saw) this maximizes your expected utility.

It is true that in both these examples, there are what we might call "extraneous" factors that are responsible for the rationality of violating conditionalization. In the first example, the violation is the only available way to attain eternal bliss; and in the second example, it is the only way to acquire some useful information. Can we show that, putting aside such considerations, it is irrational to violate conditionalization? Yes, we have already proved that. For we saw that when other things are equal (in a sense made precise in Section 5.1.6), expected utility can always be maximized without implementing a violation of Reflection. As an immediate corollary, we have that when other things are equal, expected utility can

122

always be maximized without violating conditionalization.[14]

To summarize: The principle of conditionalization is a special case of the principle that says not to implement shifts that violate Reflection. Like that more general principle, it is not a universal requirement of rationality; but it is a rationally acceptable principle in contexts where other things are equal, in the sense made precise in Section 5.1.6.

5.2.2 Other arguments for conditionalization

Lewis's Dutch book argument is not the only argument that has been advanced to show that conditionalization is a requirement of rationality. What I have said in the preceding section implies that these other arguments must also be incorrect. I will show that this is so for arguments offered by Teller, and by Howson.

After presenting Lewis's Dutch book argument, Teller (1973, 1976) proceeds to offer an argument of his own for conditionalization. The central assumption of this argument is that if you learn E and nothing else, then for all propositions A and B that entail E, if $p(A) = p(B)$, then it ought to be the case that $q(A) = q(B)$. (Here, as before, p and q are your prior and posterior probability functions, respectively.) Given this assumption, Teller is able to derive the principle of conditionalization. But the counterexamples that I have given to conditionalization are also counterexamples to Teller's assumption. To see this, consider the first counterexample, in which taking a drug will make you certain pigs can fly, and this will give you eternal bliss. Let F and d be as before, and let G denote that the moon is made of green cheese. We can suppose that in this example, $p(Fd) = p(Gd)$, and $q(Fd) = q(F) > q(G) = q(Gd)$. Assuming that d is all you learn from taking the drug, we have a violation

---

[14]Brown (1976) gives a direct proof of a less general version of this result. What makes his result less general is that it applies only to cases where for each E you might learn, there is a probability function q such that you are sure q would be your probability function if you learned E, and nothing else. This means that Brown's result is not applicable to the coin-tossing example of Section 5.1.2, for example. (In this example, your posterior probability, on learning that you took the drug, could give probability 1 to either heads or tails.) Another difference between Brown's proof and mine is that his does not apply to probability kinematics (cf. Section 5.3).

123

of Teller's principle. But the shift from p to q involves no failure of rationality. You do not want $q(F)$ to stay small, or else you will forgo eternal bliss; nor is there any reason to become certain of G, and preserve Teller's principle that way. Thus Teller's principle is not a universal requirement of rationality, and hence his argument fails to show that conditionalization is such a requirement. (My second counterexample to conditionalization could be used to give a parallel argument for this conclusion.)

Perhaps Teller did not intend his principle to apply to the sorts of cases considered in my counterexamples. If so, there may be no dispute between us, since I have agreed that conditionalization is rational when other things are equal. But then I would say that Teller's defense of conditionalization is incomplete, because he gives no method for distinguishing the circumstances in which his principle applies. By contrast, the decision-theoretic approach I have used makes it a straightforward matter of calculation to determine under what circumstances rationality requires conditionalization.

I turn now to Howson's argument for conditionalization (Howson and Urbach 1989, pp. 67f.). Howson interprets $p(H)$ as the betting quotient on H that you now regard as fair, $p(H|E)$ as the betting quotient that you now think would be fair were you to learn E (and nothing else), and $q(H)$ as the betting quotient that you will in fact regard as fair after learning E (and nothing else). His argument is the following. (I have changed the notation.)

$p(H|E)$ is, as far as you are concerned, just what the fair betting-quotient would be on H were E to be accepted as true. Hence from the knowledge that E is true you should infer (and it is an inference endorsed by the standard analyses of subjunctive conditionals) that the fair betting quotient on H is equal to $p(H|E)$. But the fair betting quotient on H after E is known is by definition $q(H)$.

I would not endorse Howson's conception of conditional probability. But even granting Howson this conception, his argument is fallacious. Howson's argument rests on an assumption of the following form: People who accept "If A then B" are obliged by logic to accept B if they learn A. But this is a mistake; on learning A you might well decide to abandon the conditional "If A then B," thereby preserving logical consistency in a different way.

124

In the case at hand, Howson's conception of conditional probability says that you accept the conditional "If I were to learn E and nothing else, then the fair betting quotient for H would be $p(H|E)$." Howson wants to conclude from this that if you do learn E and nothing else, then logic obliges you to accept that the fair betting quotient for H is $p(H|E)$. But as we have seen, this does not follow; for you may reject the conditional. In fact, if you adopt a posterior probability function q, then your conditional probability for H becomes $q(H|E) = q(H)$; and according to Howson, this means you now accept the conditional "If I were to learn E and nothing else, then the fair betting quotient for H would be $q(H)$." In cases where conditionalization is violated, $q(H) \neq p(H|E)$, and so the conditional you now accept differs from the one you accepted before learning E.

Thus neither Teller's argument nor Howson's refutes my claim that it is sometimes rational to violate conditionalization. And neither is a substitute for my argument that, when other things are equal, rationality never requires violating conditionalization.

5.2.3   Van Fraassen on conditionalization

In a recent article, van Fraassen (in press) argues that conditionalization is not a requirement of rationality. From the perspective of this chapter, that looks at first sight to be a paradoxical position for him to take. I have argued that conditionalization is a special case of the principle not to implement shifts that violate Reflection. If this is accepted, then van Fraassen's claim that Reflection is a requirement of rationality implies that conditionalization is also a requirement of rationality.

I think the contradiction here is merely apparent. Van Fraassen's idea of how you could rationally violate conditionalization is that you might think that when you get some evidence and deliberate about it, you could have some unpredictable insight that will cause your posterior probability to differ from your prior conditioned on the evidence. Now I would say that if you satisfy Reflection, your unpredictable insight will be part of what you learned from this experience, and there is no violation of conditionalization. But there is a violation of what we could call

Evidence-conditionalization. *If your current probability function is p, and if q is the probability function you would have if you acquired evidence E and no other evidence, then q(·) should be identical to p(·|E).*

This principle differs from conditionalization as I defined it, in having E be the total *evidence* acquired, rather than the totality of what was *learned*. These are different things because, as argued in Section 5.1.5, not all learning involves getting evidence. Where ambiguity might otherwise arise, we could call conditionalization as I defined it *learning-conditionalization*.

These two senses of conditionalization are not usually distinguished in discussions of Bayesian learning theory, presumably because those discussions tend to focus on situations in which it is assumed that the only learning that will occur is due to acquisition of evidence. But once we consider the possibility of learning without acquisition of evidence, evidence-conditionalization becomes a very implausible principle. For example, suppose you were to think about the value of $\sqrt{289}$, and that as a result you substantially increase your probability that it is 17. We can suppose that you acquired no evidence over this time, in which case evidence-conditionalization would require your probability function to remain unchanged. Hence if evidence-conditionalization were a correct principle, you would have been irrational to engage in this ratiocination. This is a plainly false conclusion. (On the other hand, there need be no violation of learning-conditionalization; you may think you *learned* that $\sqrt{289}$ is 17.)

So van Fraassen is right to reject evidence-conditionalization, and doing so is not inconsistent with his endorsement of Reflection. But that endorsement of Reflection does commit him to learning-conditionalization; and I have urged that this principle should also be rejected.

5.2.4 The rationality of arbitrary shifts

Speaking of the theory of subjective probability, Henry Kyburg writes:

126

But the really serious problem is that there is nothing in the theory that says that a person should *change* his beliefs in response to evidence in accordance with Bayes' theorem. On the contrary, the whole thrust of the subjectivist theory is to claim that the history of the individual's beliefs is irrelevant to their rationality: all that counts at a given time is that they conform to the requirements of coherence. It is certainly not required that the person got to the state he is in by applying Bayes' theorem to the coherent degrees of belief he had in some previous state. No more, then, is it required that a rational individual pass from his present coherent state to a new coherent state by conditionalization.... For all the subjectivist theory has to say, he may with equal justification pass from one coherent state to another by free association, reading tea-leaves, or consulting his parrot. (Kyburg 1978, pp. 176–7)

The standard Bayesian response to this objection is to claim that conditionalization has been shown to be a requirement of rationality, for example, by the diachronic Dutch book argument (Skyrms 1990b, ch. 5). But I have shown that the arguments for conditionalization are fallacious and that the principle is not a general requirement of rationality. Nevertheless, Kyburg's objection is still mistaken.

If you think there is something wrong with revising your probabilities by free association, reading tea-leaves, or consulting your parrot, then presumably shifts in probability induced by these means do not satisfy Reflection for you. If that is so, then the theorem of Section 5.1.6 shows that if these shifts would make any difference at all to your expected utility, then implementing them would not maximize expected utility, other things being equal. Thus under fairly weak conditions, Bayesian theory does imply that it is irrational for you to revise your beliefs by free association, and so forth.

5.3 PROBABILITY KINEMATICS

It is possible for the shift from p to q to satisfy Reflection without it being the case that there is a proposition E such that $q(\cdot) = p(\cdot|E)$. When this happens, you think you have learned something, but there is no proposition E that expresses what you learned. The principle of conditionalization is then not applicable.

127

Jeffrey (1965, ch. 11) proposed a generalization of conditionalization, called probability kinematics, that applies in such cases. Jeffrey supposed that what was learned can be represented as a shift in the probability of the elements of some partition $\{E_i\}$. The rule of probability kinematics then specifies that the posterior probability function q be related to the prior probability p by the condition

$$q(\cdot) = \sum_i p(\cdot|E_i)q(E_i).$$

Armendt (1980) has given a Dutch book argument to show that the rule of probability kinematics is a requirement of rationality. But this argument has the same fallacy as the Dutch book arguments for Reflection and conditionalization. Furthermore, my account of the true status of conditionalization also extends immediately to probability kinematics.

A natural interpretation of what it means for you to think what you learned is represented by a shift from p to q' on the E_i would be that the shift is to q, and

$$p(\cdot|R_q) = \sum_i p(\cdot|E_i)q'(E_i).$$

But then it follows that the requirement to update your beliefs by probability kinematics is equivalent to the requirement not to implement any shifts that violate Reflection. Hence updating by probability kinematics is not in general a requirement of rationality, though it is a rational principle when other things are equal, in the sense of Section 5.1.6.

5.4 CONCLUSION

If diachronic Dutch book arguments were sound, then Reflection, conditionalization, and probability kinematics would all be requirements of rationality. But these arguments are fallacious, and in fact none of these three principles is a general requirement of rationality. Nevertheless, there is some truth to the idea that these three principles are requirements of rationality. Bayesian decision theory entails that when other things are

128

equal, rationality never requires implementing a shift in probability that violates Reflection. Conditionalization and probability kinematics are special cases of the principle not to implement shifts that violate Reflection. Hence we also have that when other things are equal, it is always rationally permissible, and may be obligatory, to conform to conditionalization and probability kinematics.

6

The concept of acceptance

6.1 DEFINITION

In everyday life, and also in science, opinions are often expressed by making categorical assertions, rather than by citing personal probabilities. The categorical assertion of H, when sincere and intentional, expresses a mental state of the assertor that I refer to as *acceptance* of H. This chapter will examine the concept of acceptance just defined; the following chapter will argue that it is an important concept for the philosophy of science.

What I am here calling *acceptance* is commonly called *belief*. Consider, for example, G. E. Moore's (1942, p. 543) observation that it is paradoxical to say 'H, but I do not believe that H.' The paradox is explained by the fact that sincere and intentional assertion of H is taken to be a sufficient condition for believing H.

It is, then, *part* of the folk concept of belief that it is a mental state expressed by sincere, intentional assertions. However, I think the folk concept involves other aspects too, and I am going to argue that the various aspects of the folk concept of belief do not all refer to the same thing. That is why I am calling the concept I have defined 'acceptance', rather than 'belief'.

My definition of acceptance assumes that you understand what a sincere assertion is. If that is not so, we are in trouble. For I cannot define sincerity except by saying that an assertion is sincere iff it was intended to express something the assertor accepts; and you won't understand this definition unless you already understand acceptance. But I think that everyone knows what counts as evidence for and against the sincerity of an assertion, and this shows that everyone has the sort of understanding of sincerity that we require here. (Note that my aim here is not to reduce the concept of acceptance to other concepts, but merely to identify what I mean by 'acceptance'. Thus the fact that the notion of sincerity cannot be explained without invoking the notion of acceptance does not mean the definition of acceptance cannot achieve its purpose.)

I say that acceptance is expressed by assertions that are both sincere and intentional. What I mean by an intentional assertion is that it asserts what the person intended to assert; thus slips of the tongue are not intentional assertions. Unintentional assertions may very well be sincere; sincerity requires only the intention to assert something that is accepted and not that this intention is successful. Thus you may sincerely utter something you do not accept; however, you cannot make a sincere and *intentional* assertion of something you do not accept, as acceptance is here understood.

There is no infallible test for whether or not a person accepts *H*. For one thing, there is no infallible test of whether an assertion is sincere and intentional. And even if we had an infallible test of sincerity and intentionality, that test would not determine whether a person who does not assert *H* accepts *H*. But verificationism is now in disrepute, so I suppose I do not have to argue that a concept can be legitimate, even though there is no infallible test for its application. What is more important is that we be able to make reasonable inductive inferences about what a person accepts; this we can do, as the following examples will illustrate.

Categorical assertions predominate in science textbooks. Because these assertions appear to be sincere, and because they usually do not appear to be slips, we can say that the authors of science texts accept the assertions that occur in these books. Presumably their colleagues would also sincerely endorse most of these assertions, for example in teaching; and so they too accept most of what is in textbooks of their subject.[1]

In their research articles, scientists often mix categorical assertions with more guarded statements. Here is an example,

---

[1] Some texts present an outmoded theory (e.g., classical mechanics), because it is easier for students than the theory that is actually accepted (e.g., relativity theory). Often this is done by describing the theory, without asserting it. Where the outmoded theory is asserted, the assertion must be deemed insincere, even if pedagogically sound.

taken almost at random, from an article in a recent issue of *Nature* (15 June 1989 issue).

We found that this enzyme RNA catalyses the site-specific attack of guanosine on the isolated $P1$ stem, but that the K_m for free $P1$ was very high ($> 0.1mM$). This weak interaction probably reflects the fact that there are few sequence or size requirements for the recognition of $P1$ by the core intron.

Here is another example, from the same issue of *Nature*.

In conclusion, Greenland ice cores reveal abrupt and radical changes in the North Atlantic region during the Younger Dryas–Pre-Boreal transition, including decreased storminess, a 50% increase in the precipitation rate, a 7°C warming, and probably a temporary decrease of the evaporation temperature in the source area of moisture ultimately precipitated as snow at high elevation in the arctic.

In each of these quotations, categorical assertions are followed by statements that do not make a categorical assertion but rather express a judgment of probability. From these statements we can reasonably infer that the authors accept the assertions they have made categorically, but we cannot conclude that they accept the hypotheses they merely say to be probable. When scientists refrain from categorically asserting a hypothesis in a publication, this does not necessarily mean that they do not accept the hypothesis. They may accept the hypothesis themselves, but refrain from categorically asserting it in a publication because the evidence is not yet enough to convince the scientific community at large. In such a case, the scientists may say in private that they believe the hypothesis, and this would be strong evidence that they accept it. The point is that if there is any context in which a person would assert a hypothesis, then we can reasonably infer that the person accepts that hypothesis (provided it is reasonable to think that the person is sincere and not making a mistake in that context, and that the change of context would not change what the person accepts).

Acceptance of H is the state expressed by sincere intentional assertion of H. But what sort of thing is H here? Although assertions are made by means of sentences, I do not intend that H should be understood to be a sentence. I would say that what a German speaker asserts by saying 'Der Schnee ist weiss' is the same as what an English speaker asserts by saying 'Snow is white', though the sentences are different. I would also say that the state of acceptance these assertions express (if sincere and intentional) is the same. Conversely, a sentence like 'This is an electron' may be used to make quite different assertions on different occasions (possibly assertions with different truth values), even though the sentence is the same; here the state of acceptance that the assertions express (if sincere and intentional) is also different.

I therefore take H to be a proposition, rather than a sentence. Propositions are here understood as sets of states (they are also referred to as events). Thus we identify what a person has asserted on a given occasion with the set of states that are consistent with the assertion. For example, an utterance of 'Snow is white' (or 'Der Schnee ist weiss') asserts that the true state is a member of the class of all states in which snow is white. And an utterance of 'This is an electron' asserts that the true state is a member of that class of states in which the thing denoted by 'this' is an electron.

6.2 ACCEPTANCE AND PROBABILITY

6.2.1 Probability 1 not necessary

What is the relation between acceptance and probability? One suggestion would be to identify acceptance of a hypothesis with assignment of probability 1 to that hypothesis. But this view is untenable. For to give hypothesis H probability 1 is to be willing to bet on it at any odds; for example, a person who gave H probability 1 would be willing to accept a bet in which the person wins a penny if H is true, and dies a horrible death if H is false. I think it is clear that scientists are not usually this confident of the hypotheses they sincerely categorically assert, and thus that probability 1 is not a necessary condition for acceptance.[2]

---

[2] I argued that this is so in (Maher 1986b) and (1990c, n. 13). Because I think few readers will need convincing, I do not repeat those arguments here.

133

6.2.2 High probability not sufficient

Having discarded the idea that acceptance can be identified with probability 1, we might try identifying it with "high" probability, that is, probability greater than (or not less than) some value $r < 1$. But this also would be a mistake. For example, consider a lotto game in which six numbers between 1 and 50 are drawn without replacement. If you purchase a ticket in such a game, your chance of winning is less than one in ten million. Now suppose you are thinking of purchasing such a ticket; you have selected six numbers, and you ask me whether I think this combination will win. I certainly would not say it *will* win, but I also would not say it *won't* win. I would tell you the odds against it winning are enormous, but no greater than for any other number. Nor would I be being less than forthcoming here; I simply would not accept the proposition that your numbers will not win, even though I give this proposition an enormously high probability. Hence even an extremely high probability is not sufficient for acceptance.

The claim of the preceding paragraph is that it is logically possible to give a proposition an extremely high probability, yet not accept it. Even if this is conceded, it might still be claimed that it is irrational not to accept a proposition with an extremely high probability. But I think this claim should also be rejected. I do not think the stance I adopted in the preceding paragraph is irrational. Furthermore, the claim conflicts with what I take to be a more compelling principle of rationality.

If I were to accept everything that I think extremely probable, then for every set of six lotto numbers, I would accept that this set would not win. However, I also accept that some set of six lotto numbers would win. Thus I accept an inconsistent set of propositions. Hence if rationality required me to accept all the propositions I regard as highly probable, it would require me to accept an inconsistent set of propositions. But we naturally suppose that rationality requires consistency.

That natural supposition needs some qualification to be defensible. It may be that there is some nonobvious inconsistency in the propositions I accept, but the only way for me to remove it would be to spend years investigating the logical relations among the things I accept, or else to give up much of what I

134

accept. Either way, the cost of achieving consistency seems more than it is worth, and so it is rational for me to continue with the inconsistency.

However, the lotto example I was discussing is not like this. Here I have a clearly identified inconsistency, which can easily be removed. In such a situation, I think rationality requires consistency. Our everyday practice suggests that this endorsement of consistency is widely shared. Show people that various things they have said are inconsistent, and they will feel obliged to retract one of those propositions. But if rationality requires consistency, even if only in cases where inconsistency is easily avoidable, then rationality cannot require accepting all propositions with high probability.[3]

6.2.3 Probability 1 not sufficient

I have argued that no probability short of 1 is sufficient for acceptance, while a probability of 1 is not necessary for acceptance. Together, these results show that acceptance cannot be identified with any level of probability.

Still, it may seem that if a proposition is given a probability of 1, then it must be accepted. But this also is false. Suppose Professor Milli's probabilities concerning the charge of the electron, e, are as depicted in Figure 6.1. Here $f(e)$ denotes Milli's "probability density" for e. What this means is that for any numbers a and b, Milli's probability that e lies between a and b is the area under the curve $f(e)$ from a to b. As we can see from the figure, for any distinct values a and b, Milli gives a positive probability to e being between a and b. However, for any number a, Milli's probability that $e = a$ is zero. (The "area" under the curve $f(e)$ from a to a has zero width, and hence is zero.) Since Milli's probabilities satisfy the axioms of probability, it follows that for every number a, Milli gives probability 1

---

[3]The clash of principles here is the core of the "lottery paradox", which seems to have been introduced to the literature by Kyburg (1961, p. 197). Kyburg took the opposite stance to the one I am taking; he upheld the principle that high probability is sufficient for acceptance, and so rejected consistency. The difference between him and me may derive from the fact that his conception of acceptance differs from mine, being tied to practical action. For a critique of Jonathan Cohen's (1977) attempt to resolve the paradox with "inductive" probabilities, see Maher (1986b, p. 375f.).

135

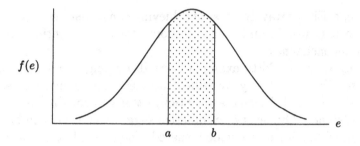

Figure 6.1: Milli's probability distribution for e

to the proposition that $e \neq a$. So if probability 1 were sufficient for acceptance, then for each number a, Milli must accept that $e \neq a$.

But suppose Milli says: "My conclusion is that the charge on the electron is $4.774 \pm .009 \times 10^{-10}$ electrostatic units." Pressed further, he refuses to rule out any value of e in this interval. Then absent any indications that Milli is being less than forthcoming, I think we should allow that for values of a in the stated interval, Milli does not accept that $e \neq a$. And this need not undermine our attribution to Milli of the probability density function f, that gives probability 1 to all propositions of the form $e \neq a$. Thus probability 1 is not sufficient for acceptance.

If you have gone this far with me, you concede that it is logically possible for a person to give a proposition probability 1, yet not accept it. This still leaves open the possibility that probability 1 rationally obliges one to accept a proposition, even if it does not logically force it. But I think there is also no such rational obligation. Indeed, the idea that there is such an obligation, like the parallel idea considered in the preceding section, conflicts with the principle of consistency.

The conflict is evident in Milli's case. If Milli were to accept every proposition to which he gives probability 1, then for all a, he would accept that $e \neq a$. Yet Milli is also sure that the electron has some value or other, and hence must accept this too. Thus if probability 1 is sufficient for acceptance to be rationally required, Milli is rationally obliged to accept an inconsistent set of propositions. Yet there is nothing pathological about Milli's

136

probability function; probability functions like this are called "normal" in statistics. The problem lies rather in the idea that probability 1 is sufficient for rational acceptance.

6.2.4 High probability not necessary

For all that I have said so far, it could be that high probability is necessary for acceptance. Specifically, it may be thought that a proposition cannot be accepted without giving it a probability of at least 1/2. While I would agree that accepted propositions commonly do have a probability greater than 1/2, I do not think that this must always be the case.

It has often been noted that, in view of the regularity with which past scientific theories have been overthrown, we can reasonably infer that current scientific theories will also be overthrown. Thus anyone reflecting on the history of science ought to give a low probability (less than 1/2) to any given significant current theory being literally correct. Yet scientists continue to sincerely assert significant scientific theories. If high probability were necessary for acceptance, we would have to say that either

(a) These scientists have not drawn the appropriate lesson from the history of science, and accord their theories an unreasonably high probability; or

(b) Contrary to appearances, these scientists do not actually accept their theories.

While (a) might be true in many cases, we could have good evidence that it is false for some scientists. Suppose Einstein were offered his choice of

(1) World peace if general relativity is completely correct; or

(2) World peace if general relativity is false in some way.

I think Einstein probably would have chosen (2). In any case, let us suppose that he makes this choice. Then it would be reasonable to conclude (using the preference interpretation of probability) that his probability for general relativity is less than 1/2.

If high probability is necessary for acceptance, we then would have to say that Einstein did not accept general relativity. But suppose he does categorically assert the theory, at the same time that he is choosing (2) over (1); this is certainly possible. It is also possible that we could satisfy ourselves that no slip of the tongue occurred; his assertion was intentional. Then someone who holds that high probability is necessary for acceptance must say that Einstein's assertion is not sincere. "Actions speak louder than words," it will be said.

But suppose Einstein defends himself against this charge of insincerity. "General relativity is simple in conception, follows from attractive assumptions, and its empirical predictions have been successful. Thus although the theory is probably incorrect in some way, I am confident that it will live on as a limiting case of a future theory. And in the meantime, it is the only theory that fits the evidence. Thus for the time being I accept that the world is as the theory says, though I realize that the theory is likely to be corrected in the future." With such an explanation, and absent any other reasons to question sincerity, I think we should accept that Einstein is sincere in asserting general relativity, even though giving it a low probability of being completely correct. If someone thinks otherwise, their conception of sincerity is different from mine.

There may be a temptation to say that what Einstein really accepts is not general relativity itself, but rather the proposition that general relativity is *approximately* correct; that it will "live on as a limiting case of a future theory." But I am supposing that, as scientists usually do, Einstein is categorically asserting the theory itself, not merely that the theory is approximately correct. That being so, what he accepts, on my definition of acceptance, is the theory itself, not merely the claim that the theory is approximately correct.

The example I have been using is largely fictitious, though I hope it is not too implausible a fiction. In any case, the mere possibility of the story is enough to establish that it is possible to accept a proposition without giving it a probability as high as 1/2.

But even if acceptance is possible in this case, can it be rational? I submit that the example at hand supports an affirmative answer. It is certainly rationally permissible (if not obligatory) to give major scientific theories a low probability of being literally correct. But we also pretheoretically suppose that it is rational to accept our best current scientific theories. If these things seem in conflict, I would suggest the cause is an inadequate theory of rational acceptance. In the next section, I will describe a theory that lets us say all the things we want to say here.

We don't need to go to grand scientific theories to find examples in which acceptance of propositions, while giving them probability less than 1/2, can be rational. I think that for almost everyone, there are ten propositions that the person accepts, but to whose conjunction the person would give probability less than 1/2. The propositions can be mundane things like 'My desk is made of oak', 'My wife is in the living room', 'Tom is a psychology major', and so on. In such cases, giving the conjunction a probability less than 1/2 seems reasonable; this follows from the assumption that we are not much more than 90 percent confident of each, and their probabilities are independent. Accepting the individual propositions is also pretheoretically reasonable. So someone who maintains that a probability greater than 1/2 is necessary for rational acceptance must say that rationality does not require people to accept the logical consequences of what they accept. But this logical principle is one on which we rely every day, so abandoning it is a high price to pay.

6.3 RATIONAL ACCEPTANCE

6.3.1 Theory

I have argued that high probability is neither necessary nor sufficient for rational acceptance of a hypothesis. If this is right, rational acceptance must depend on something other than probability.

To see what this something else is, consider the conclusion Cavendish drew from an experiment he conducted in 1773. The experiment was to determine how the electrostatic force between charged particles varies with the distance between the particles. Cavendish states his conclusion this way:

We may therefore conclude that the electric attraction and repulsion must be inversely as some power of the distance between that of the $2 + 1/50$th and that of the $2 - 1/50$th, and there is no reason to think

139

that it differs at all from the inverse duplicate ratio. (Cavendish 1879, pp. 111–2)

This statement indicates that Cavendish accepted

H_C: The electrostatic force falls off as the nth power of the distance, for some n between 1.98 and 2.02.

Why wouldn't Cavendish have accepted only a weaker conclusion, for example by broadening the range of possible values of n, as in

H_C': The electrostatic force falls off as the nth power of the distance, for some n between 1.9 and 2.1.

Or he could have made his conclusion conditional, as in

H_C'': If the electrostatic force falls off as the nth power of the distance, for some n, then n is between 1.98 and 2.02.

Both H_C' and H_C'' are more probable than the conclusion H_C that Cavendish actually drew, as are infinitely many other weaker versions of Cavendish's hypothesis. The obvious suggestion is that although these weaker hypotheses are more probable than H_C, they are also considerably less informative, and that is why Cavendish did not limit himself to these weaker hypotheses. But if informativeness is what is wanted, why not accept a stronger hypothesis, a natural one here being

H_C''': The electrostatic force falls off as the second power of the distance.

And again there is an obvious answer: Although H_C''' is more informative than H_C, Cavendish felt that it was not sufficiently probable to accept.

These considerations suggest that acceptance involves a trade-off of two competing considerations: the concern to be right (which would lead one to accept hypotheses of high probability), and the desire for informative hypotheses (which tends to favor hypotheses of low probability). Thus a theory of acceptance needs to take into account the scientist's goals or values, and specifically the relative weights put on the goals of truth and informativeness.

140

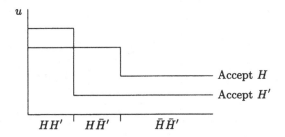

Figure 6.2: A simple conception of the utility of accepting hypotheses

Since the Bayesian way to represent goals is with a utility function, a Bayesian theory of acceptance thus requires there to be a utility function representing the weights that the scientist puts on the competing goals of truth and informativeness.

What would such a utility function be like? First we need to identify the consequences that are the domain of the utility function. A simple suggestion (Hempel 1960, 1962; Levi 1967) is that acceptance of a hypothesis H has two possible consequences, which we could describe as "accepting H when it is true," and "accepting H when it is false." The utility function would assign utilities to consequences such as these, and the goal of truth would be represented by giving a higher utility to the former consequence than to the latter. Then the utility of accepting H would be as in Figure 6.2.

If H' is a logically stronger hypothesis than H, then the goal of accepting *informative* true hypotheses would be represented by giving higher utility to the consequence "accepting H' when it is true" than to "accepting H when it is true." This is also diagrammed in Figure 6.2. (In this figure, the utility of accepting a false hypothesis is represented as being higher for the less informative hypothesis; but I do not insist that this must be the case.)

Figure 6.2 assumes that the utility of accepting a given hypothesis depends only on the truth value of the hypothesis, and thus has only two possible values. But it has often been said that false theories can be more or less close to the truth; or as Popper puts it, some false theories have greater *verisimilitude* than others. And it is held that the utility of accepting

141

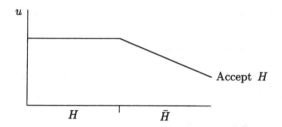

Figure 6.3: Utility of acceptance depending on verisimilitude

a false theory depends on this distance from the truth, being higher the closer the hypothesis is to the truth. On this view, not only will the utility of accepting a hypothesis depend on whether the hypothesis is true or false, but also, in case the hypothesis is false, it will depend on how far from the truth the hypothesis is. So a plot of the utility of acceptance might look like Figure 6.3.

Up to this point I have been discussing the acceptance of a single hypothesis. But scientists accept many hypotheses, and the utility of accepting two hypotheses will not in general be the sum of the utilities of accepting each. However, we can reduce the general case to the case of accepting a single proposition by focusing on the total corpus of accepted propositions. For this corpus can be represented by a single proposition K, namely the conjunction of all propositions that the person accepts. The acceptance of a new proposition H, without any change in those previously accepted, can then be represented as a replacement of the total corpus K by the logically stronger corpus consisting of the conjunction of K and H. The abandonment of a previously accepted proposition, and the replacement of one accepted proposition by another, can also be represented as shifts from one corpus to another. From this holistic perspective, a scientist's goals can be represented by assigning utilities to the possible consequences of accepting K as corpus, for each possible corpus K. I will refer to a function of this sort as a *cognitive utility function*, since it assigns utilities to cognitive consequences.

Chapter 8 will show how a cognitive utility function can be defined, consistently with the preference interpretation of probability and utility. In the meantime, I will anticipate that result

142

and assume that a cognitive utility function has been defined. We can then define the *expected cognitive utility* of accepting a corpus in the usual way, as the probability-weighted sum of the utilities of the possible consequences. So if $u(K, x)$ denotes the cognitive utility of accepting K as corpus in state x, and if the set of all possible states is X, then the expected cognitive utility of accepting K as corpus is $\sum_{x \in X} p(x)u(K, x)$, assuming X is countable. We can then say that acceptance of corpus K is rational just in case the expected cognitive utility of accepting this corpus is at least as great as that of accepting any other available corpus.

6.3.2 Example

To illustrate this theory of rational acceptance, consider the problem of estimating the true value of some real-valued parameter. For example, the problem might be to estimate the true value of the charge on the electron. The problem can be phrased, in our terminology, as: For what set A of real numbers should I accept (as corpus) that the true value of the parameter is in A? Since the hypotheses in which we are interested differ only in the set A, it will be convenient to identify the set with the corresponding proposition.

Let us define the *content* of (the proposition that the true value is in) A as[4]

$$
c(A) = \begin{cases} \frac{1}{1 + \sup(A) - \inf(A)} & \text{if } A \neq \emptyset \\ 1 & \text{if } A = \emptyset. \end{cases}
$$

So, for example, the content of a set containing just a single point is 1, and it declines as the set A is enlarged, reaching 0 when A is the whole real line $(-\infty, \infty)$. Also, let A's *distance from the truth*, when the true value of the parameter is r, be defined as

$$
d_r(A) = \begin{cases} \frac{\inf_{x \in A} |x - r|}{1 + \inf_{x \in A} |x - r|} & \text{if } A \neq \emptyset \\ 1 & \text{if } A = \emptyset. \end{cases}
$$

[4]Here $\sup(A)$ denotes the least upper bound, or supremum, of A; and $\inf(A)$ denotes the greatest lower bound, or infimum, of A. If A is the interval (a, b) or $[a, b]$, then $\sup(A) = b$ and $\inf(A) = a$.

143

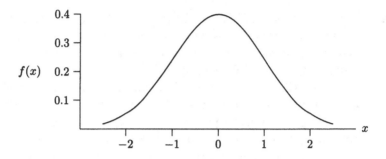

Figure 6.4: The standard normal distribution

If A is true (i.e., $r \in A$), then $d_r(A) = 0$. As the distance between r and the closest point in A increases, $d_r(A)$ increases, approaching 1 in the limit.

Now for any real number k, define

$$u_k(A, r) = kc(A) - d_r(A).$$

For each k, u_k represents a possible measure of the cognitive utility of accepting A when the true state is r; and k represents the relative weight this utility function puts on the desideratum of informativeness, as compared with the competing desideratum of avoidance of error. I believe that there are other possible cognitive utility functions besides the u_k; I define these functions here merely to give an example.

To complete the stipulations of this example, suppose that some scientist's probability distribution for the parameter is the standard normal distribution; this means that for any set A, the probability that the true value r is in A equals the area above the set A under the curve $f(x)$ in Figure 6.4.[5] Milli might have this distribution if r is a multiple of $e - 4.774 \times 10^{-10}$.

If cognitive utility is measured by u_k, for some $k < 1$, then expected cognitive utility can always be maximized by accepting as corpus a closed interval of the form $[-a, a]$, for some a.

[5]For those who want the formula, it is

$$p(A) = \frac{1}{\sqrt{2\pi}} \int_A e^{-x^2/2} \, dx.$$

The set A must be measurable.

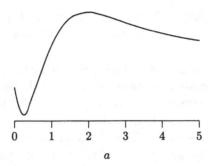

Figure 6.5: $\mathcal{E}(u_{.37}[-a,a])$ plotted as a function of a

Proof. For any $A \neq \emptyset$, the interval $[\inf(A), \sup(A)]$ has the same content as A, and cannot have a larger error. Also,

$$u_k(\emptyset, r) = k - 1 < 0 = u_k((-\infty, \infty), r),$$

and so expected utility is not maximized by accepting $\emptyset$. Thus expected utility can always be maximized by accepting a corpus that is a closed interval. Since the standard normal distribution is symmetric about 0, the closed intervals that maximize expected utility have the form $[-a, a]$.

Thus we lose no real generality if we assume that the sets A that are candidates for acceptance are all closed intervals of the form $[-a, a]$.

Figure 6.5 shows the expected utility of accepting $[-a, a]$, plotted as a function of a, when $k = .37$.[6] Expected utility is maximized at $a = 1.95$, and thus it is rational to accept as corpus that the true value of the parameter is in the interval $[-1.95, 1.95]$. The probability that the true value is indeed in

[6]I use $\mathcal{E}$ to denote expected value. The formula is

$$
\begin{aligned}
\mathcal{E}(u_k[-a,a]) &= \frac{1}{\sqrt{2\pi}} \int_{-\infty}^{\infty} u_k([-a,a], x) e^{-x^2/2} \, dx \\
&= \frac{k}{1+2a} - \sqrt{\frac{2}{\pi}} \int_{a}^{\infty} \frac{x-a}{1+x-a} e^{-x^2/2} \, dx.
\end{aligned}
$$

145

this interval is .95, and thus the interval $[-1.95, 1.95]$ can be thought of as the Bayesian analog of a 95 percent confidence interval.

Note that in accepting $[-1.95, 1.95]$ as corpus, one suspends judgment on propositions that are more probable than what is accepted. For example, to accept $[-1.95, 1.95]$ as corpus is to suspend judgment on whether the value is outside the interval $[-.01, .01]$; and the probability of the latter is .994. Despite its higher probability, acceptance of the latter proposition (either as corpus, or in conjunction with $[-1.95, 1.95]$) would reduce expected cognitive utility. Similarly for the negation of even smaller intervals, which have even higher probability. In accepting $[-1.95, 1.95]$, one also fails to accept that the value of the parameter is not precisely 0, though that proposition has probability 1. Unlike the interval cases, addition of this proposition to $[-1.95, 1.95]$ would not *reduce* expected utility, because it does not reduce the content of what is accepted or increase the possible error; however, it does not increase expected utility either, so there is no positive reason to accept it. Furthermore, since the set of all propositions with probability 1 is inconsistent, to accept all propositions with probability 1 is to accept the empty set $\emptyset$ as corpus, which does not maximize expected utility (see preceding proof). Thus the present theory of rational acceptance supports my earlier claim that no level of probability is sufficient for acceptance.

As the value of k is increased, the cognitive utility function u_k puts more weight on content. Figure 6.6 shows $\mathcal{E}(u_{.5}[-a, a])$, plotted as a function of a, assuming again that the parameter has a standard normal probability distribution. Here the maximum expected utility is attained when $a = 0$, and thus expected utility is maximized by taking as corpus the degenerate interval $[-0, 0]$, that is, the singleton set $\{0\}$. Since this set has probability 0, one sees that *expected cognitive utility can be maximized by accepting a proposition with zero probability*. The reason is that although there is no chance of the corpus being literally correct, there is a very good probability that it will be quite close to the truth, and this together with the desire for an informative corpus makes acceptance of the corpus optimal. This dramatically supports my earlier intuitive argument, that

146

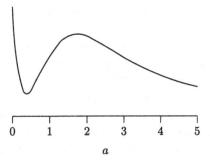

Figure 6.6: $\mathcal{E}(u_{.5}[-a, a])$ plotted as a function of a

high probability is not a necessary condition for accepting a hypothesis.[7]

6.3.3 Objections

Epistemologists have stock objections to decision-theoretic accounts of rational acceptance. One is that we don't have probabilities and utilities; the answer to this is in Sections 1.3 and 1.5. Another is that expected utility calculations are difficult to do; the answer to this is in Section 1.2. Here I wish to answer two other objections that are often raised, and have not been addressed earlier in this book.

The first of these objections is this: The acts that are evaluated in decision theory must be options that you could choose if you wanted to. But a doxastic state like acceptance is not something you can choose at will, like deciding to take an umbrella. Hence the application of decision theory to acceptance is fundamentally misguided.

To see the answer to this objection, we need to think about the role norms of rationality play. These norms get their point because our acceptance of them tends to influence what we do. Thus there is no point having norms requiring us to do f, if f is so far beyond our power that acceptance of this norm could have no tendency to make us do f. For example, it would be

---

[7]Niiniluoto (1986) gives a similar Bayesian account of interval and point estimation.

147

futile for me to accept the norm 'Have blue eyes,' given that doing so would have no tendency to make me have blue eyes. Typically, the things that acceptance of a norm cannot get us to do are the things that are not subject to our will; and hence we get the principle that 'ought' implies 'can'. However, the case of acceptance is not typical in this respect. While acceptance is not usually directly subject to the will, it nevertheless is the case that acceptance of norms governing acceptance influences what propositions we accept. We see this every day in science, for example where scientists determine a 95 percent confidence interval and then accept that the true value of the variable of interest lies in this interval. If the scientists accepted a different normative statistical theory, they would draw a different conclusion. Hence there is a point to saying that we *ought* to accept this or that proposition, even if acceptance is not directly subject to the will.

Thus for the purposes of cognitive decision theory, we take the options to be not alternatives that are directly subject to the will but alternatives that we would (or could) take if our norms required this. It seems that we could accept any proposition we can formulate, if we accepted a norm requiring this. Hence there is no shortage of options to which to apply cognitive decision theory.

The second objection I wish to address is this: Suppose acceptance of A maximizes expected utility, and that I do accept it. Then according to the theory I have outlined, I have done the rational thing. But suppose that I accepted A for some irrelevant reason; perhaps I just picked it randomly and it was a fluke that I happened to pick the proposition that maximizes expected cognitive utility. Then, the objection claims, my acceptance of A was irrational, even though it maximizes expected utility.

This objection has an air of unreality about it, since I am not likely to be able to accept something for a reason that I think is irrelevant. And if we want to make the case one in which I do not think my reason is irrelevant, but ought to, then my irrationality can be attributed to having irrational probabilities and/or utilities (Section 1.9).

But putting aside such cavils, let us suppose that I can choose what to accept by a process that I regard as random. I hold

148

that this would not be irrational unless the random process had some positive chance of resulting in me accepting something with lower expected utility than A. But in that case, the expected utility of randomly choosing what to accept is less than the expected utility of directly accepting A. This is so even if, as luck would have it, the random method results in me accepting A. For the expected utility of randomly deciding what to accept is a mixture of the expected utilities of the various propositions that I might accept by this method, and in the case we are dealing with, this must be less than the expected utility of A. Thus my theory of rational acceptance adequately explains the irrationality involved in accepting A as the result of a random process.

6.4 ACCEPTANCE AND ACTION

My definition of acceptance linked acceptance of H to assertion of H. Is there also a relation between acceptance and action? Only an indirect one, I shall argue.

Note first that acceptance of H does not commit you to act in all cases as if H were true. For to act as if H were true is to act in the way that would be optimal if H were indeed true (where what counts as "optimal" is determined by your utilities). Thus if you were committed to acting in all cases as if H were true, you would be committed to accepting any bet on H, since you will win such a bet if H is true. But I have already noted that you can accept a proposition without being willing to bet on it at any odds.

Conversely, you can be willing to act in all cases as if H were true yet not accept H. For we have seen that you can give H a probability of 1 yet not accept H; and if you give H a probability of 1, you are willing to act in all cases as if H were true.

Nor is it the case that acceptance of H commits you to betting on H when the odds exceed some threshold. For to be willing to bet on H at odds greater than $m : n$ is to have a probability for H greater than $m/(m + n)$; and we have seen that no positive level of probability is necessary for acceptance.

There is thus no necessary link between what a person accepts and how the person acts in practical circumstances. In any situation, acceptance of H is consistent both with acting

149

as if H were true and also with acting as if H were false, and nonacceptance of H is also compatible with acting either way. This reflects the fact that rational action is determined by probabilities (plus utilities), and acceptance is not identifiable with any level of probability.

The lack of linkage between acceptance and action may also be brought out by imagining two rational individuals who have the same probability distributions, and whose utility functions agree on practical consequences, but who assign different utilities to some cognitive consequences. Then both individuals will have identical preferences regarding all practical actions but may nevertheless accept different propositions. Conversely, if their utility functions agree on cognitive but not practical consequences, they will accept the same propositions but have different preferences regarding practical actions.

Furthermore, the decision to accept a theory (or to not accept it) normally should produce no change in one's willingness to act as if the theory were true in practical contexts. If I rationally accept H, then H maximizes expected cognitive utility, relative to my current probability and cognitive utility functions. The fact that I have accepted H would not normally give me any reason to further increase the probability of H. For example, the reasons that make me confident the theory of evolution is true do not include the fact that I accept the theory. Thus the decision to accept a theory (or to not accept it) normally should produce no change in the probability of that theory – and hence no change in one's willingness to act as if the theory were true in practical contexts.

If this were not so (i.e., if the acceptance of a theory produced a change in one's willingness to act in accordance with the theory), then acceptance of a theory would have an influence on practical utilities, and this would need to be taken into account in considering whether or not it is rational to accept the hypothesis. The fact that there is normally no such influence means that, normally, practical utilities are irrelevant to the rationality of accepting a hypothesis. That is why, in presenting my theory of acceptance in Section 6.3, I supposed that rational acceptance maximizes expected *cognitive* utility, and ignored practical utilities.

150

Having just mentioned my acceptance of the theory of evolution, let me indicate what role acceptance has played in our evolution. I have been arguing that acceptance of a hypothesis normally has no influence on practical action; and our survival, insofar as it is up to us, is completely determined by our practical actions. Why then would evolution produce creatures like us, who tend to place considerable importance on accepting truths, and avoiding accepting falsehoods? The reason, I suggest, is that our intellectual curiosity provides us with a motivation to reason, observe, and experiment, and these activities often tend ultimately to favor our practical success. The desire for practical success can also motivate such information gathering, but intellectual curiosity provides an additional, and often more immediate, stimulus.

This does *not* mean that intellectual curiosity is reducible to a desire for practical success, and hence that cognitive utility is a species of practical utility. Since the point is somewhat subtle, let me illustrate it with an unsubtle analogy: the desire for sex. It is clear that we have this desire because individuals who have it are more likely to reproduce. However, the desire for sex is not the same as a desire to reproduce, or else there would be no market for contraceptives. I am suggesting that intellectual curiosity (cognitive utility) is like the desire for sex: Just as the desire for sex increases our motivation to reproduce, so our intellectual curiosity increases our motivation to acquire information; and both of these tend to increase evolutionary fitness – but what the desire is a desire for is not the same as the evolutionary function that the desire serves. It may be that astrophysics is the intellectual equivalent of a contraceptive: something that breaks the connection between our desires and the evolutionary function these desires serve.

Although there is no necessary linkage between acceptance and practical action, we may sometimes be able to make inferences about what people accept on the basis of their practical actions. For people's practical actions give us some information about their probabilities; and from that, together with some assumptions about their cognitive utilities, we can make an inference about what they accept. This is the same process we use to predict people's actions in one practical context, on the basis of their actions in a different practical context.

There is, then, some connection between acceptance and practical action, due to the fact that both partially reflect a person's probability function. But this indirect connection is the only connection there is.

6.5 BELIEF

At the beginning of this chapter, I mentioned my reservation about the folk concept of belief. We have now reached the point where I can explain the problem with this concept.

It is standardly assumed that you believe H just in case you are willing to act as if H were true. But under what circumstances is this willingness to act supposed to occur? As we observed in the preceding section, to be committed to *always* acting as if H were true is to be willing to bet on H at any odds. However, the usual view seems to be that you do not need to be absolutely certain of H (give it probability 1) in order to believe it. For one thing, it is usually supposed that there is very little we can be rationally certain of, but that we can nevertheless rationally hold beliefs on a wide range of topics. Thus belief in H does not seem to imply a willingness to act as if H were true under all circumstances.

Two responses to this difficulty suggest themselves. One is to say that you believe H just in case you are willing to act as if H were true, *provided the odds are not too high* (where what counts as "too high" remains to be specified). On this account, belief in H would be identified with having a probability for H exceeding some threshold. The other approach is to abandon the idea that belief is a qualitative state that you either have or you don't, and instead say that it comes in degrees. Your degree of belief in H could then be measured by the highest odds at which you would be willing to bet on H. This second suggestion effectively identifies belief with probability; and in fact, Bayesians since Ramsey (1926) have referred to subjective probability as "degree of belief."

Whichever of these suggestions is adopted, one will still have to deal with another aspect of the concept of belief. As I

152

remarked at the beginning of this chapter, it is standardly assumed that belief in H is the mental state expressed by sincere intentional assertion of H – in other words, that belief is the same as acceptance. But if belief is acceptance, then it cannot be related to probability in either of the ways just considered.

The reason why acceptance cannot be identified with probability greater than some threshold has already been given in Section 6.2. No matter where the threshold is set, you can give H a probability that high but still not accept H; and you also can accept something with a probability lower than the threshold. The reason why acceptance cannot be identified with probability itself is that acceptance (as I have defined it) is a qualitative state that one either has or lacks, while probability is a matter of degree. One might think of dealing with the latter problem by introducing a notion of "degree of acceptance," and attempting to identify that with probability; but this also will not work. On any reasonable understanding of degree of acceptance, a person who (qualitatively) accepts H but not K, ought to count as having a higher degree of acceptance for H than for K; yet we have seen that such a person may well give a lower probability to H than to K.

The upshot, then, is that the folk concept of belief appears to regard belief in H as a single mental state that is expressed both by a willingness to act as if H were true and also by sincere intentional assertion of H; and these are in fact two distinct states.

Stich (1983, pp. 230–7) has claimed that the mental states underlying sincere assertion and practical action need not be the same, and has inferred that the folk concept of belief may not refer to anything. Obviously I agree with Stich that these states need not be the same; for I have argued that they *are* *not* the same.

Stich cites some psychological research that, he claims, supports the conclusion that the mental states underlying assertion and action are distinct. I would like to be able to cite some empirical support of this kind, for the conclusion I have reached on more a priori grounds. But unfortunately, the research cited by Stich gives no real support to this conclusion (though it also does not disconfirm it).

Consider, for example, a study by Wilson, Hull, and Johnson (1981), discussed by Stich. In this study, subjects were induced to volunteer to visit senior citizens in a nursing home, and then later were asked if they would volunteer to help former mental patients. For one group of subjects, overt pressure from the experimenter to visit the senior citizens was made salient; for another group, it was not. There was a negative correlation between being in the first group and agreeing to help the mental patients. The experimenters hypothesized that the reason for the correlation was that subjects in the second group, feeling less pressured, would tend to infer that they were visiting the nursing home because they were helpful people, and that this inference made them more likely to volunteer to help the mental patients.

After agreeing to visit the nursing home, half the subjects in each group were asked to list all the reasons they could think of that might explain why they agreed to go and to rate their importance. Later, all subjects were asked to rate themselves on various traits relevant to helpfulness. Those who had been asked to list reasons tended to rate themselves as more helpful than those who had not, but they were not more likely to volunteer to help the mental patients.

This study does show that subjects' actual and self-reported helpfulness can be independently manipulated. But it does not follow from this that different types of belief state underlie assertion and action; for subjects who act in helpful ways are not acting on the belief that they are helpful. What follows is rather that *being* helpful and accepting that you are helpful are two different states.

A suitable experiment to show the difference between action-producing and assertion-producing belief states would be to have subjects consider a million-ticket fair lottery. If they are not willing to assert categorically that a given ticket will not win, but are willing to bet on this at high odds, then we have established that (a high degree of) the kind of belief that underlies action is not sufficient for the kind of belief that underlies assertion. Conversely, if we can find propositions that subjects will categorically assert to be true but will not bet high odds on, we have shown that the kind of belief that underlies

154

assertion is not sufficient for (a high degree of) the kind of belief that underlies action. I have already suggested that, at least for reflective people, scientific theories often fall in the latter category.

6.6 OTHER CONCEPTS OF ACCEPTANCE

The term 'acceptance' is widely used, especially in the philosophy of science; and other authors have given it definitions that differ from the one I have given in this chapter. I will conclude this chapter by comparing my definition with a few of these alternatives.

6.6.1 Kaplan

Mark Kaplan (1981) has advocated a conception of acceptance that is similar in spirit to mine. Like me, Kaplan views rational acceptance as maximizing expected cognitive utility, and he asserts that improbable propositions can be rationally accepted, while probable ones need not be. But he and I define acceptance in different ways.

Kaplan defines 'S accepts H' as meaning

S would defend H were S's sole aim to defend the truth.

By 'defend' Kaplan means categorical assertion; so his definition and mine are related in connecting acceptance to assertion. Also, if S's sole aim were to defend the truth, then S would be sincere; and so the notion of sincerity figures in both definitions.

The main difference between Kaplan's definition and mine is that his definition uses a counterfactual conditional. This results in his definition's having an extension different from mine. For example, on Kaplan's account, only very confused persons could ever accept

A: Defending the truth is not my sole aim.

For if your sole aim were to defend the truth, and you were not confused, you would know A was false, and hence would not defend it. On my account, you could accept A because you could be in the mental state expressed by sincere intentional assertion of A (even though you might not wish to actually assert it).

155

So far this difference is just a different choice of how to use a word. But when combined with the decision-theoretic account of rational acceptance that Kaplan and I share, it becomes a substantive difference. Suppose that you are in the mental state expressed by sincere intentional assertion of A, but would not defend A were your sole aim to defend the truth, since then you would know A was false. My theory says that you accept A, and hence that your present cognitive utility is higher if A is true than if A is false. Kaplan's theory says that you accept $\bar{A}$, and hence that your present cognitive utility is higher if A is false than if A is true. I think my theory better fits our sense of what your cognitive utility would be in this situation.

Another difference is that Kaplan does not take account of the possibility of unintentional assertions. Suppose that if your sole aim were to defend the truth, then you would attempt to assert H, but you would make a slip, and unintentionally assert H'. Then on Kaplan's account you accept H', but on my account you probably[8] accept H. So on Kaplan's account, your cognitive utility, if you would slip and assert H', is the same as if you would sincerely and intentionally assert H'; while on my account, it is the same as if you were to sincerely and intentionally assert H. Again, I think my account better fits our sense of what your cognitive utility would be in this situation.

6.6.2 Levi

Levi (1967) discussed two notions of acceptance, which he called 'acceptance as true' and 'acceptance as evidence'. Levi (1967, p. 25) says that 'accept as true' means the same as 'believe'. However, Levi allows that you can accept H as true, without being willing to act on H in all contexts; he even allows that you can accept H while giving it a low probability. Thus acceptance-as-true departs from the common concept of belief by not being linked with action. Levi does seem to assume, however, that sincere intentional assertion of H expresses a person's acceptance of H as true (1967, pp. 10, 223). In both these

---

[8]If you are attempting to assert H, and if you are sincere, then on my account you accept H. But in the case we are considering here, this only shows that if your sole aim were to defend the truth, then you would (on my account) accept H. It remains possible that you do not in fact accept H.

156

respects, acceptance as I have defined it appears to agree with Levi's concept of acceptance-as-true.

Levi also offered a decision-theoretic account of when it is rational to accept a hypothesis as true. Like the account of rational acceptance I have sketched, Levi's account proposed that the cognitive goals of accuracy and content could be represented by a cognitive utility function. However, Levi held that a person might simultaneously have different demands for information, these demands being representable by different utility functions. And for Levi, acceptance-as-true was relative to these demands for information; on his theory, one could accept H as true to satisfy one demand for information, and at the same time accept $\bar{H}$ as true to satisfy another demand for information. This would be rational, according to Levi, provided each acceptance-as-true maximized expected cognitive utility, relative to its own particular cognitive utility function.

My notion of acceptance differs from Levi's concept of acceptance-as-true in not being relative to demands for information. My main reason for not adopting a relativized notion is that rationality requires keeping the various propositions we accept consistent with one another; thus someone who accepts H to satisfy one of their demands for information is obliged to also accept H as an answer to any other demand for information that H may satisfy. We therefore do not need a question-relative notion of acceptance in a theory of rational acceptance.

In Levi's later writings (1976, 1980), the notion of acceptance-as-true has virtually disappeared, and acceptance-as-evidence occupies center stage. It is now acceptance-as-evidence for which Levi gives a decision-theoretic account involving cognitive utility. I have the impression that after (1967), Levi came to view acceptance-as-true as not a real bearer of cognitive utility. On this later view, as I interpret it, to accept H as true relative to some demand for information is merely to be committed to accept H as evidence if (possibly contrary to fact) that demand for information were one's only demand.[9] This interpretation of Levi implies that he too does not now believe that acceptance, in the sense that influences

---

[9]This interpretation is suggested by (Levi 1976, p. 35, and 1984, p. xiv).

cognitive utility, is relative to demands for information.

But while Levi's notion of acceptance-as-evidence is like my notion of acceptance in not being relativized, it differs from my notion in another crucial respect: To accept a hypothesis as evidence, Levi says, is to assume its truth in all (practical and theoretical) deliberation. Thus if one accepts H as evidence, one must give H probability 1. By contrast, my notion of acceptance, like Levi's notion of acceptance-as-true, allows that one can accept hypotheses without giving them probability 1, and indeed can do so while giving them a probability as low as you like.

I find the concept of acceptance-as-evidence less important than Levi does, because I think people accept-as-evidence much less than he supposes. Levi imagines that observation reports, scientific theories, and other things are often accepted-as-evidence. This means that people should be willing to bet at any odds on a wide range of observation reports and scientific theories. I am not willing to do that, and people I have talked to say they are not willing either. I do not deny that some propositions, even contingent ones, are assigned probability 1; if one is dealing with a real-valued parameter, for example, some contingent propositions must get probability 1. But I take it that the propositions given probability 1 are typically very uninformative ones, like "the force between charged particles does not vary precisely as the rth power of the distance." Observation reports and scientific theories are typically regarded as fallible.

In any event, acceptance as I have defined it is ubiquitous, and is not identifiable with acceptance-as-evidence. For as Levi has observed (1967, p. 10), people can and do sincerely assert propositions they are not willing to bet on at all odds and hence do not take to be certainly true; the propositions thus asserted are accepted in my sense, but not accepted-as-evidence. Thus a theory of acceptance in my sense is needed, whatever one thinks of Levi's concept of acceptance-as-evidence.

6.6.3 Van Fraassen

In The Scientific Image, van Fraassen maintained that acceptance of a theory in science involves belief that the theory is empirically adequate (it agrees with observable phenomena) but not that the theory is true. He also said that this belief in empirical adequacy is only a necessary condition for acceptance, not a sufficient condition. Acceptance, he said, also involves a certain commitment. For scientists, the relevant commitment entails the adoption of a particular kind of research program; and for nonscientists, it entails "willingness to answer questions *ex cathedra*" (van Fraassen 1980, p. 12).

For van Fraassen (1985, pp. 247ff.), belief is the same as probability. So his statement, that acceptance of a theory in science involves belief in its empirical adequacy, must mean that acceptance of a theory in science involves giving a high probability to the theory being empirically adequate. And similarly, the statement that belief in the truth of the theory is not necessary for acceptance in science must mean that one can accept a theory in science without giving a high probability to the theory being true.

Van Fraassen's reference to ex cathedra pronouncements points to a similarity between his concept of acceptance and mine: Both are connected with categorical assertion. However, the connection is less direct on my account. People, including scientists, are sometimes secretive; they can be in the mental state expressed by sincere intentional assertion of H, without being willing to assert H. Thus I would not agree that acceptance entails "willingness to answer questions *ex cathedra*."

On both van Fraassen's view and mine, a person can accept a theory without giving a high probability to the theory being true. However, I disagree with his statement that acceptance of a theory in science involves belief that the theory is empirically adequate. Van Fraassen's reason for saying this is that he thinks (a) science aims at accepting empirically adequate theories, and (b) (rational?) acceptance involves a high probability that the aim of acceptance is served. Proposition (a) is about the aim of science, not the concept of acceptance; I will discuss it in Section 9.6. Here I focus on (b), which is a claim about the nature of (rational) acceptance.

One problem with (b) – and also with (a) – is that the notion of an aim seems to presuppose that there is only one desirable outcome; whereas in fact there may be a continuum of possible outcomes, of varying degrees of desirability. But let us put that problem aside, and suppose that we have a simple case in which we can say that the aim in accepting H is to accept an empirically adequate proposition. Still, rational acceptance of H does not require that H have a high probability of being empirically adequate. For example, suppose that accepting H increases utility by 10 if H is empirically adequate, and reduces utility by 1 otherwise. Then the probability of H being empirically adequate need only be 0.1 for acceptance of H to be rational. Thus (b) is false.

Let us turn now to the question of whether acceptance of a hypothesis commits a scientist to a research program. Van Fraassen says that this commitment to a research program comes about because accepted theories are never complete (1980, p. 12); this suggests that the research program that he takes to be entailed by acceptance is one of extending the scope of the theory. But most scientists extend the scope of only a few theories, if any, and I doubt that van Fraassen wishes to keep the class of theories accepted by scientists this small. Perhaps the commitment to extend the scope of the theory is meant to apply only to those scientists who extend the scope of some theory in the subject area of the accepted theory: The position would be that if you accept H, then if you extend the scope of some theory in the subject area of H, you must extend H. But even this is not a condition that must hold for acceptance as I have defined it. For example, we can easily imagine a scientist Poisson', who sincerely asserted a corpuscular theory of light, but extended the scope of Fresnel's wave theory of light by deriving a previously unnoticed consequence of the theory, and extended the scope of no other theory of light.[10] Poisson' would accept a corpuscular theory on my account of acceptance but not on van Fraassen's account.

Even if your sole concern was the scientific one of accepting the right theory, it could still be rational to accept (in my sense) one theory but work on an incompatible theory. To see this, suppose that acceptance of H currently has a higher expected cognitive utility than does the acceptance of H', but there is

---

[10]The real Poisson did extend the scope of Fresnel's theory, without accepting it (Worrall 1989). But I do not know whether Poisson accepted the corpuscular theory, or whether he extended the scope of that theory.

a chance H' could be developed further into a theory whose acceptance would have a higher expected utility than H. In such a case, it could be rational to develop H' while accepting H.[11]

6.7 SUMMARY

I defined acceptance of H as the mental state expressed by sincere intentional assertion of H. We saw that whether a person will accept H depends not only on the probability of H but also on other factors, such as how informative H is. In the light of this, I proposed a decision-theoretic account of rational acceptance, in which acceptance is viewed as having consequences with cognitive utility, and rational acceptance maximizes expected cognitive utility.

Acceptance, as I have defined it, captures one aspect of the folk notion of belief, while probability is a different, incompatible aspect of that concept. This notion of acceptance is related to the notions of acceptance employed by Kaplan, Levi, and van Fraassen; where it differs from the latter, I have given reasons for preferring my definition.

Two main questions remain to be answered. One is why (if at all) we should think that acceptance is an important concept for the philosophy of science. The other is the question of how to justify the assumption of a utility function for cognitive consequences. These questions will be considered, in this order, in the next two chapters.

---

[11] Essentially this point has been made by Laudan (1977, pp. 108–13). However, the point seems to be inconsistent with Laudan's notion of acceptance, which is *treating a theory as if it were true* (p. 108).

7

The significance of acceptance

In the previous chapter, I argued that the doxastic state of accepting a hypothesis is not reducible to probability, and I sketched a Bayesian theory of rational acceptance that takes into account cognitive utilities as well as probabilities. But a theory of acceptance is not a usual component of Bayesian philosophies of science. It seems that many Bayesian philosophers of science think of subjective probability as a replacement for the notion of acceptance, and so think that acceptance has no important role to play in a Bayesian philosophy of science. This chapter will argue that that view is a mistake, by describing three ways in which the theory of acceptance makes an important contribution to Bayesian philosophy of science.

7.1 EXPLAINING THE HISTORY OF SCIENCE

Much of what is recorded in the history of science is categorical assertions by scientists of one or another hypothesis, together with reasons adduced in support of those hypotheses and against competing hypotheses. It is much less common for history to record scientists' probabilities. Thus philosophers of science without a theory of acceptance lack the theoretical resources to discuss the rationality (or irrationality) of most of the judgments recorded in the history of science. But a philosophy of science this limited in scope can fairly be described as impoverished.

Without a theory of acceptance, it is also impossible to infer anything about scientists' subjective probabilities from their categorical assertions. Thus for a philosophy of science without a theory of acceptance, the subjective probabilities of most scientists must be largely inscrutable. This severely restricts the degree to which Bayesian confirmation theory can be shown to agree with pretheoretically correct judgments of confirmation that scientists have made. But as we saw in Section 4.2, demonstration of such agreement provides an important argument for the correctness of Bayesian confirmation theory. Thus the lack of a theory of acceptance would seriously limit the arguments supporting the correctness of Bayesian confirmation theory.

This conclusion will be surprising to most Bayesian philosophers of science. For while Bayesian philosophers of science do not generally see a need for a theory of acceptance, they have nevertheless gone to considerable lengths to analyze episodes in the history of science. How is this possible, when according to what I have just said, analysis of the history of science normally requires a theory of acceptance? It is possible because although these philosophers do not see the need for a theory of acceptance, they nevertheless operate with a tacit theory of acceptance. This tacit theory seems to identify acceptance with high probability. In Section 6.2 I explained why this is an untenable theory. Consequently, Bayesian analyses of historical episodes do not avoid reliance on a theory of acceptance; and they could be improved by using a better theory of acceptance.

To illustrate this, I will review three representative examples of Bayesian attempts to explain episodes from the history of science, and show how each of them presupposes an incorrect theory of acceptance.

One reason I used Cavendish as an example in Section 6.3.1 is that this example has been analyzed by Dorling (1974). Dorling takes the fact that needs explaining to be that "Cavendish's experiment and argument ... render it highly probable that the correct law would be at any rate a very good macroscopic approximation to the inverse-square law" (1974, p. 336); and he gives a Bayesian analysis that yields this conclusion. But this analysis explains a historical fact only if Cavendish (or other scientists) did think his experiment made it highly probable that an inverse square law was close to the truth. And history does not record Cavendish saying that he thought this. It simply records his categorical conclusion that the electrostatic law was approximately an inverse-square law.[1] Nor does it seem that

---

[1] For the relevant quotation from Cavendish, see page 139.

other scientists said Cavendish's experiment made his conclusion highly probable; certainly Dorling cites no evidence that they did.

Presumably Dorling took Cavendish's assertion to show that Cavendish gave his conclusion a high probability. Dorling would then be tacitly assuming

(1) Accepted propositions are given high probability.

If my argument in Section 6.2.4 is right, then (1) is false, and so Dorling's analysis tacitly presupposes an untenable theory of acceptance. This is not to say that Dorling's analysis is fatally flawed; perhaps the theory of acceptance developed in Chapter 6 could be used to argue that Cavendish did indeed give a high probability to his conclusion, as a result of his experiment. But that is not my concern here. Rather my concern is simply to point out that Dorling's historical analysis does tacitly presuppose a theory of acceptance and that the analysis would be improved by using a better theory of acceptance.

My next example is a historical analysis by Franklin and Howson (1985), who offer what they say is a Bayesian explication of Newton's argument from Kepler's laws to the inverse-square law of gravitation. Newton gives this argument in his *Principia*. As Franklin and Howson point out, Newton offers a deductive argument from Kepler's laws, together with a result stated earlier in the *Principia*, to the conclusion that the inverse-square law holds between each planet and the sun.[2] Newton also presents evidence that the inverse-square law holds between the moon and the earth, and between Jupiter and Saturn and their moons. Franklin and Howson's explanation of why Newton offers this additional evidence is that "The additional evidence supports the hypothesis that the inverse square force is universal."

Since Franklin and Howson claim to be giving a Bayesian explication, the notion of "support" they invoke here needs a Bayesian interpretation. I guess that what they mean by "support" is confirmation, or increase in probability. Thus if M is

[2] This derivation has been criticized because (a) it assumes that the sun is fixed in space, contrary to what Newtonian theory implies, and (b) the earlier result Newton uses for the derivation applies to circular motion only, not the elliptical orbit of the planets. Franklin and Howson argue that Newton was aware of these facts but regarded them as introducing negligible errors.

164

the evidence about the moons, and U the universal inverse-square law, Franklin and Howson's explication of why Newton offers M is that $p(U|M) > p(U)$. And in fact, it can easily be argued that this inequality should hold. If the existence of the moons and planets is assumed, then $p(M|U) = 1$; and provided $p(M) < 1$, the desired inequality then follows from Bayes' theorem.

But what reason is there to think that this analysis explicates Newton's reasoning? Newton did not say that $p(U|M) > p(U)$; rather he lists M as a premise in his argument for the categorical conclusion U. If Franklin and Howson are attempting to explain Newton's probability judgments, they must be inferring what those judgments were from Newton's use of this argument, in which case they are probably assuming

(2) If evidence strengthens the case for accepting a hypothesis, then it confirms that hypothesis.

Alternatively, Franklin and Howson might be trying to explain why Newton argued as he did, by showing that Newton would have held $p(U|M) > p(U)$. In that case, they would seem to be assuming

(3) If evidence confirms a hypothesis, then it strengthens the case for accepting that hypothesis.[3]

However, both principles (2) and (3) are false.

For example, suppose we have 100 exclusive and exhaustive hypotheses, denoted H_1 through H_{100}. If their prior probabilities are all the same, namely 0.01, then we would most likely suspend judgment, and not accept or reject any of them. Suppose this is so. Now we acquire some evidence E that raises the probability of H_1 to 0.98, and the probability of H_2 to 0.02, while the probabilities of the remaining hypotheses drop to zero. With the new probabilities, we would most likely accept H_1. In that case, E has confirmed H_2 (it has raised its probability from 0.01 to 0.02), yet has led to H_2 being rejected and so certainly

---

[3]This principle has been explicitly endorsed by van Fraassen, who asserts that "any reason for belief is a fortiori a reason for acceptance" (1983, p. 168). Van Fraassen identifies belief with (high) subjective probability.

165

has not strengthened the case for accepting H_2. Thus we have a counterexample to (3). Also, E has led to $\bar{H}_2$ being accepted and so has strengthened the case for accepting $\bar{H}_2$, yet E disconfirmed $\bar{H}_2$ (it reduced the probability of $\bar{H}_2$ from 0.99 to 0.98); this is a counterexample to (2).

This example is consistent with the decision-theoretic account of acceptance given in Section 6.3.1.

Proof. For $1 \leq n \leq 100$, let the utility of accepting a disjunction of n of the H_i be $\frac{100}{n} - 1$ if the disjunction is true, and -2 if the disjunction is false. Then expected utility is initially maximized by accepting the disjunction of all 100 H_i, that is, suspending judgment. And after E is learned, expected utility is maximized by accepting H_1 alone.

Thus Franklin and Howson's explication of Newton's reasoning tacitly presupposes a false principle regarding acceptance. I think this flaw in their explication could be repaired by a more careful analysis, but that is not my concern here. The point I wish to make is merely that their analysis does not avoid reliance on a theory of acceptance; rather it tacitly assumes part of a theory, and what they assume does not stand up to critical scrutiny. Their explication of Newton's reasoning could be improved by the use of a better theory of acceptance.

For my third and last cautionary tale, I turn to Rosenkrantz's (1977) analysis of Copernicus's evidence for the heliocentric theory. Rosenkrantz writes:

[The] simplifications of the Copernican system (by virtue of which it is a *system*) are frequently cited reasons for preferring it to the Ptolemaic theory. Yet, writers from the time of Copernicus to our own, have uniformly failed to analyse these simplifications or account adequately for their force. The Bayesian analysis ... fills this lacuna in earlier accounts by showing that the cited simplifications render the heliostatic theory ... *better supported* than the corresponding (more complicated) geostatic theory. (1977, p. 140)

Let H be Copernicus's heliocentric hypothesis, G Ptolemy's geocentric hypothesis, and E the evidence of the observed planetary motions. Rosenkrantz's definition of support is such

166

that E supports H better than G just in case

$$\frac{p(H|E)}{p(H)} > \frac{p(G|E)}{p(G)}.$$

As the preceding quote indicates, Rosenkrantz intends this inequality to be a reason for preferring H to G.

Now history records that Copernicus accepted H and rejected G, and it records the reasons he gave in support of this, but it does not record his probabilities for H or G. Thus if Rosenkrantz's account is to explain the historical facts, the preference that it explains must be a preference for *accepting* H rather than G. Rosenkrantz's account would then be tacitly assuming

(4) If E supports H better than G, then E is a reason to prefer accepting H over accepting G.

Alternatively, we might interpret Rosenkrantz as inferring from the historical record that Copernicus regarded E as supporting H better than G (in Rosenkrantz's sense) and attempting to explain why this probabilistic judgment should hold. In that case, his inference from the historical record appears to assume

(5) If E is a reason to prefer accepting H over accepting G, then E supports H better than G.

However, both principles (4) and (5) are also false.

For example, suppose A, B, and C are mutually exclusive and exhaustive hypotheses, with probabilities before and after acquiring evidence E as shown in Figure 7.1. One might well suspend judgment on these three hypotheses before learning E, and accept A afterward. In that case, E is clearly not a reason to prefer accepting C over accepting A; nevertheless, E does support C better than A. Thus we have a counterexample to (4). Also, E is a reason to prefer accepting A over C, though it does not support A better than C; so we also have a counterexample to (5). I leave it to the interested reader to show that this example is consistent with the theory of Section 6.3.1.

The preceding example illustrates a situation that is ubiquitous in science. For example, let A be some hypothesis that is

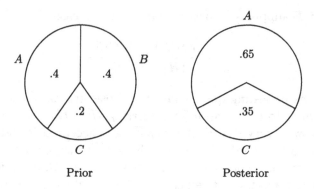

Prior Posterior

Figure 7.1: Counterexample to (4) and (5)

confirmed by evidence E, but does not logically entail E; and let C be the proposition $\bar{A}E$. Then

$$\frac{p(A|E)}{p(A)} = \frac{p(E|A)}{p(E)} < \frac{1}{p(E)} = \frac{p(E|C)}{p(E)} = \frac{p(C|E)}{p(C)}.$$

Thus whenever a hypothesis A does not logically entail evidence E, the proposition $\bar{A}E$ is better supported, in Rosenkrantz's sense, than is A; but it would be absurd to infer from this that E makes acceptance of $\bar{A}E$ preferable to acceptance of A. There are many cases in which we take evidence to provide a good reason for accepting a hypothesis that does not entail that evidence.

Thus Rosenkrantz's explanation of Copernicus's reasoning, like the two earlier explanations I have considered, relies on a tacit theory of acceptance that does not stand up to critical scrutiny. To make his explanation cogent, a better theory of acceptance must be employed.

With these three examples, I have illustrated the point that I made at the beginning of this section, namely: A theory of acceptance is normally required to show that historical judgments of confirmation by scientists can be explained by Bayesian confirmation theory. Since the ability to explain exemplary episodes in the history of science can provide important support for Bayesian confirmation theory, this is one reason why even Bayesians who find acceptance intrinsically uninteresting

168

should nevertheless regard the theory of acceptance as important. More reasons follow.

7.2 THE ROLE OF ALTERNATIVE HYPOTHESES

In *The Structure of Scientific Revolutions*, Kuhn wrote:

[O]nce it has achieved the status of paradigm, a scientific theory is declared invalid only if an alternate candidate is available to take its place. No process yet disclosed by the historical study of scientific development at all resembles the methodological stereotype of falsification by direct comparison with nature.... The decision to reject one paradigm is always simultaneously the decision to accept another, and the judgment leading to that decision involves the comparison of both paradigms with nature *and* with each other. (p. 77)

I think that what Kuhn means by a "paradigm" here is a theory that has been so successful as to be regarded as an exemplary achievement. And so understood, Kuhn's statement makes an observation that appears to be substantially correct and important for understanding the development of science. A good example, and one cited by Kuhn, is provided by the history of Newtonian gravitational theory. From around 1690 to 1750 it was well known that the moon's perigee precessed at twice the rate calculated by Newton himself, but this did not cause Newton's theory to be rejected, and in the end the discrepancy turned out to be due to a mathematical error. Again, in 1859 Le Verrier published his calculations showing that according to Newtonian theory, the precession of the perihelion of Mercury should be about 7 percent less than the observed value. The latter anomaly never was satisfactorily resolved within Newtonian theory and was part of the evidence that supported Einstein's relativity theory. But Newton's theory continued to be accepted in the intervening period despite the outstanding anomaly, and the lack of a better alternative surely was an important reason for this continued acceptance.

Of course, the advance of the perihelion of Mercury was not the only evidence supporting Einstein's theory, and it by itself may not have been enough to persuade many scientists to reject Newton's theory in favor of Einstein's. Other important evidence supporting Einstein was the null result of the Michelson–Morley experiment, and the results of Eddington's

1919 measurement of the curvature of light passing the sun. But the Michelson–Morley result was also known before Einstein's theory was proposed, and had not led to the rejection of Newton's theory. One can only speculate on what would have happened if Eddington's results had been obtained before Einstein's theory was proposed, but I think that this too would not have been enough to lead many scientists to reject Newton's theory in the absence of Einstein's theory explaining the results. Eddington's results were not completely unambiguous and were obtained under difficult conditions; they could easily have been attributed to experimental error. One might also suspect some error in the assumptions on which the result was based, such as the mass of the sun. If this is right, then the evidence that in fact persuaded many scientists of the falsity of Newton's theory would not have done so if Einstein's alternative theory had not been available.

As this illustrates, contrary evidence is rarely enough to persuade scientists to reject a theory that has been highly successful; only when there is also an alternative theory that accounts for the evidence better is the older theory likely to be rejected. This is a fact that is important for understanding the development of science, and it is something that a satisfactory philosophy of science should be able to explain.

The decision-theoretic account of acceptance is able to explain this phenomenon in a very natural way. Suppose H is a theory that has had exemplary success (such as Newton's theory of gravitation) but faces some serious anomalies; and suppose that there is no better alternative to H available. Then the options available to scientists can be adequately represented as: to accept H, to accept $\bar{H}$, or to suspend judgment altogether (i.e., to accept only $H \vee \bar{H}$).[4] If the anomalies are sufficiently serious, $p(H)$ may be low; but since H has been highly successful, it is still probably reasonably close to the truth, since new anomalies do not negate earlier successes. Also H will be highly informative. By contrast, $\bar{H}$ has very little content, even though the anomalies may indicate that its probability is high, and hence its expected distance from the truth is small. And

---

[4]The symbol '$\vee$' means 'or'. Thus '$H \vee \bar{H}$' is the tautologous statement that H is either true or false.

170

	Content	Expected distance from truth
H	high	small
$\bar{H}$	low	smaller
$H \vee \bar{H}$	zero	zero

Figure 7.2: Acceptance options without an alternative theory

while $H \vee \bar{H}$ is certainly true, it has no content at all. The situation is summarized in Figure 7.2. Thus scientists who aim to accept informative theories that are close to the truth can be expected to continue to accept H, despite the anomalies.

Now suppose an alternative theory K is formulated. This gives scientists an option they did not have before, namely to accept K. We can suppose that the content of K is at least comparable with that of H (otherwise K would not really be an *alternative* to H). And if K accounts for the evidence better, or has other relevant virtues, then its expected distance from the truth will be surmised to be less than that of H. Thus accepting K will have higher expected utility than accepting H, and it is rational to accept K, thereby rejecting H (since the two are assumed to be incompatible). In this way, the formulation of the new theory leads to a rejection of the old theory, something that was not able to be effected merely by the anomalies the old theory faced.

The phenomenon to which Kuhn has drawn attention concerns the conditions under which scientists *accept* a scientific theory (or reject it – which is to accept its negation). And the explanation I have just given makes essential use of the theory of rational acceptance. To see that the notion of acceptance is essential here, consider how one might try to deal with the phenomenon without invoking acceptance.

The first difficulty is to state the phenomenon. A probabilistic ersatz version of it would be: Anomalies do not substantially disconfirm a theory in the absence of an alternative theory, but a theory can be substantially disconfirmed when an alternative is available. Now one problem with this ersatz version is that it is not borne out by the history of science, in the way that Kuhn's

171

thesis is. For the data from the history of science primarily concern the hypotheses that scientists accepted at different times, not the probabilities that scientists gave to those hypotheses. Thus even if one could give an explanation for the ersatz thesis, that would be an answer to the wrong question. What a satisfactory philosophy of science needs to be able to explain is Kuhn's thesis, and since acceptance is not high confidence, this is not explained by explaining the ersatz thesis.

Second, I see no plausible explanation for why the ersatz thesis would be true, or even any good reason to think that it is true. I'll discuss this by considering both parts of the thesis in turn: the claim that anomalies do not substantially disconfirm a well-established theory in the absence of an alternative theory, and the claim that anomalies together with an alternative theory do substantially disconfirm an established theory.

Let E be the evidence about the advance of the perihelion of Mercury obtained by the end of the nineteenth century: that its precession is 7 percent higher than predicted on the basis of Newton's theory and current views about the remainder of the solar system, that attempts to locate additional planets to account for the discrepancy have been fruitless, and so on. Surely E is much more what one would expect if Newton's theory were false, rather than true; an intuition that is supported by the fact that at this time some scientists proposed modifications to Newton's theory to account for the anomalous perihelion advance – for example, modifications to the inverse-square law. So if H is Newton's theory, we appear to have that $p(E|H) \ll p(E|\bar{H})$; but then it follows from Bayes' theorem that E substantially disconfirmed H, contrary to what the ersatz thesis asserts.

Now let D denote that an alternative theory has been developed, which better accounts for the evidence to date (e.g., Einstein's theory). The ersatz thesis has it that ED disconfirms H to a much greater extent than E alone. This will be true just in case $p(D|EH) \ll p(D|E\bar{H})$. But we know that regardless of whether H is true, there are always alternative theories that fit the data to date, so there does not seem to be any reason why $p(D|EH)$ should be much less than $p(D|E\bar{H})$; thus there seems no reason why the discovery of an alternative theory should itself substantially disconfirm H.

172

To sum up this section: Kuhn's thesis makes an important observation about the dynamics of scientific theory change, and it is one that can be naturally explained using the theory of acceptance, and cannot be explained without invoking acceptance. This is a second reason why acceptance is important in the philosophy of science.

7.3   THE SCIENTIFIC VALUE OF EVIDENCE

Evidence gathering is a central part of scientific activity, and clearly makes an important contribution to the advance of science. It is just as clear that a satisfactory philosophy of science ought to be able to explain the importance of evidence gathering in pursuing scientific goals.

But what exactly should such an explanation show? A first suggestion might be: Gathering evidence is always worth doing so far as the goals of science are concerned. But this claim is not true. Gathering evidence takes time and money, and sometimes is would be wiser not to gather a particular piece of evidence, so that the resources can be put to use on other aspects of the scientific enterprise. And even if we put aside the cost of gathering evidence, it still would not be true that there is positive value in gathering any given piece of evidence; some evidence is simply irrelevant to anything we care about. (Who cares whether the number of hairs on my head is odd or even?)

What does seem to be true is that, ignoring the costs of evidence gathering, gathering any particular piece of evidence should not be expected to be positively detrimental to our scientific goals. For ignoring costs, gathering irrelevant information neither advances nor impedes our scientific goals; and gathering relevant evidence can be expected to advance those goals, even though in some cases it might in fact retard them. For future reference, I shall state this thesis here as the

Scientific value of evidence thesis (SVET). *If there were no cost in gathering a piece of evidence, then so far as the goals of science are concerned, it would not be irrational to gather that evidence; and in some cases it would be irrational not to do so.*

173

It is the truth of SVET that a philosophy of science ought to be able to explain.

Naïvely, one might suppose that it is trivial to give such an explanation: Science aims to find true theories, and gathering cost-free evidence cannot be counterproductive so far as this goal is concerned. But this is too simplistic because it overlooks the fact that evidence is not infrequently misleading, causing scientists to accept false theories and reject true ones. For example, experiments conducted in the 1920s were taken by Bohr and other physicists to show that the principle of conservation of energy is violated in the beta decay of atomic nuclei; but we now believe that this was a mistake, resulting from ignorance of the existence of a hitherto unknown particle, the neutrino. Of course, we believe that we now have the correct interpretation, but there is no guarantee that we will always find the correct interpretation of evidence, even in the long run.

There is a result in Bayesian decision theory that has been thought to give a satisfactory explanation of SVET. I shall argue that properly interpreted this result does indeed provide a satisfactory explanation of SVET, but that the requisite interpretation involves the notion of acceptance.

The result in question has already been stated in Section 5.1.2; it is that under suitable conditions (spelled out in Section 5.1.2), gathering evidence increases the expected utility of subsequent choices, if it has any effect at all. For example, checking the weather forecast before dressing does not guarantee that I will dress appropriately for the day's weather, but it does increase the probability of this and so increases the expected utility of my subsequent decision about how to dress.

This result will explain SVET, provided we are willing to identify the scientific value of gathering evidence with the expected utility of subsequent decisions that might be made in the light of the evidence. But what are these subsequent decisions? I. J. Good (1967) appears to assume that the subsequent decisions can be identified with practical actions. He recognizes that in many cases scientists gather evidence without any definite practical applications in view, but even in such cases he supposes that the ultimate goal is still pragmatic success. He has proposed that 'pseudo-' or 'quasiutilities' be used to measure the value of these unknown practical applications (1969, p. 185) (1983, pp. 40, 191–2, 219).

Now it is true that science contributes to pragmatic success, and this provides a perfectly good motivation for doing science. But scientists also have cognitive goals, which are not reducible to pragmatic goals. This is indicated by the importance that scientists attribute to different experiments. On Good's account, the importance that scientists attribute to an experiment ought to be a function of the expected practical applications of the experiment; and this is not the case. For example, Galileo's observation of the moons of Jupiter was rightly regarded as of great scientific importance, though surely nobody expected any practical applications to flow from it, and indeed none have been forthcoming.[5]

I wish to understand the 'goals of science' referred to in SVET as being these cognitive goals. With this understanding, we cannot explain SVET as Good attempts to do, by appealing to the fact that evidence gathering increases the expected utility of subsequent practical actions.

This failure of Good's account is avoidable by making use of the notion of acceptance. We can say that the decisions influenced by evidence gathering are not limited to practical applications but may also include cognitive decisions regarding what hypothesis to accept. Our formal result then shows that gathering cost-free evidence not only increases the expected utility of practical actions, but also increases the expected utility of acceptance decisions. The utility attaching to the latter decisions reflects cognitive goals, such as the goal of accepting true informative theories. Thus with the notion of acceptance, the formal result does indeed provide an explanation of SVET.

Some authors, seeing that SVET is not explained by Good's approach, have offered an alternative approach to explaining SVET, which does not make use of the notion of acceptance. This alternative approach has been advocated, in slightly

---

[5] Of course, Galileo hoped to use his discoveries to increase his fame and his salary, and these are practical ends. But our concern here is with how evidence gathering contributes to scientific goals, and these goals of Galileo's are not scientific ones. If there were not other goals to which Galileo's discoveries contributed, then his discoveries would not have contributed to Galileo's fame or salary, either.

175

different forms, by Rosenkrantz (1981) and Horwich (1982). According to these authors, the cognitive goal of scientists is not to accept true informative theories but rather to have probabilities that are close to the truth. A probability assignment $p(H)$ is said to be close to the truth if $p(H)$ is close to 1, and H is true; or if $p(H)$ is close to 0, and H is false. On this conception, the ideal situation would be to give probability 1 to all truths and probability 0 to all falsehoods; other situations have lesser utility, depending on how closely they approximate that ideal. Call this the *probabilist* explanation of SVET, as opposed to the *acceptance-based* explanation I offered. I will now argue that the probabilist explanation is inferior to the acceptance-based explanation.

To keep things simple, suppose that all we are interested in is the truth value of some hypothesis H. Then the cognitive utility of having probability function p depends only on the truth value of H, and we can write $u(p, H)$ for the cognitive utility of p when H is true, and $u(p, \bar{H})$ for the cognitive utility of p when H is false. I will call a utility function u *truth-seeking* if for all probability functions p and p', if $p'(H) > p(H)$, then $u(p', H) > u(p, H)$, and $u(p', \bar{H}) < u(p, \bar{H})$. Now it turns out that there are truth-seeking cognitive utility functions for which SVET is false; that is, a scientist who had one of these utility functions could *reduce* expected cognitive utility by gathering cost-free evidence. For example, let

$$u(p, H) = p(H) + \frac{1}{3\pi} \sin(3\pi p(H))$$

and let $u(p, \bar{H})$ be defined simply by replacing H by $\bar{H}$ in this equation. From Figure 7.3, it can be seen that $u(p, H)$ increases as $p(H)$ increases, and thus that u is a truth-seeking cognitive utility function. Nevertheless, with this utility function, gathering cost-free evidence can reduce expected cognitive utility.

Proof. Suppose $p(H) = 0.8$, $p(E) = 0.5$, and $p(H|E) = 0.9$. The expected utility of *not* acquiring the evidence E or $\bar{E}$ is the expected utility of holding to the current probability function, namely

$$p(H)u(p, H) + p(\bar{H})u(p, \bar{H}) = 0.78$$

176

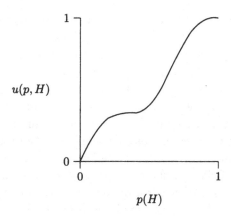

$$u(p,H)$$

$$p(H)$$

Figure 7.3: $u(p, H) = p(H) + \frac{1}{3\pi} \sin 3\pi p(H)$

Let p_E and $p_{\bar{E}}$ denote the probability functions that result from conditioning p on E and $\bar{E}$ respectively. Then the expected utility of acquiring the evidence E or $\bar{E}$ is

$$p(E)\left[p(H|E)u(p_E, H) + p(\bar{H}|E)u(p_E, \bar{H})\right] +$$
$$p(\bar{E})\left[p(H|\bar{E})u(p_{\bar{E}}, H) + p(\bar{H}|\bar{E})u(p_{\bar{E}}, \bar{H})\right] = 0.76.$$

The expected utility of gathering the evidence is thus less than that of not gathering it.

How might a defender of the probabilistic account respond to this? One possible response would be to say that as a matter of fact, scientists do not have cognitive utility functions like the one in Figure 7.3; that the cognitive utility functions scientists actually have are such as to make SVET hold. But if this is all that can be said, then SVET merely expresses a contingent fact about scientists' cognitive utility functions. I think we want to say that SVET is a deeper truth than that; that gathering cost-free evidence could never be expected to be deleterious to scientific goals.

Another possible response is to say that a utility function simply does not count as representing the cognitive goals of science unless it makes SVET true. The cognitive utility function in Figure 7.3 would, on this view, be rejected as unscientific. This

177

response avoids the objection raised in the preceding paragraph; indeed, it reduces SVET to a trivial necessary truth. But this success is only achieved at a price. The proposed restriction on what counts as a scientific utility function is not motivated by any reflection on the goals of science that the utility function is supposed to capture, but merely by the fact that the probabilist explanation fails without this restriction; thus the probabilist restriction on what counts as a scientific utility function is quite ad hoc. No such ad hocery is needed in the acceptance-based explanation of SVET; there we were able to show that SVET holds, regardless of the form of the cognitive utility function.

But even if this ad hocery is embraced, there is a further problem. Consider, for example, the cognitive utility function defined by

$$u(p, H) = p(H); \qquad u(p, \bar{H}) = p(\bar{H}).$$

Horwich (1982, pp. 127–9) postulated that this utility function represented scientific cognitive values, and showed that on this assumption SVET is deducible.[6] But if you have this utility function, then your probability function maximizes expected cognitive utility only if it assigns to H a probability of 0, 1/2, or 1; any other value is suboptimal.

Proof. Let p be your current probability for H, and let q be any number between 0 and 1. The expected cognitive utility of adopting q as the probability of H, calculated using your current probability p, is

$$\begin{aligned} \mathcal{E}_p(q) &= pu(q, H) + (1 - p)u(q, \bar{H}) \\ &= pq + (1 - p)(1 - q). \end{aligned}$$

Differentiating with respect to q gives $d\mathcal{E}_p(q)/dq = 2p - 1$. Thus

---

[6]This way of putting the matter is mine, not Horwich's. What Horwich says is that the *error* in probability function p when H is true is $1 - p(H)$, and similarly for $\bar{H}$. Since he takes rational choice to minimize expected error, he is in effect identifying cognitive utility with negative error, i.e., $u(p, H) = p(H) - 1$. Since utility functions differing by a constant represent the same preferences, it is equivalent to say $u(p, H) = p(H)$; similarly for $\bar{H}$.

if $p > 1/2$, $d\mathcal{E}_p(q)/dq$ is positive, and q maximizes expected cognitive utility iff $q = 1$. Similarly, if $p < 1/2$ then q maximizes expected cognitive utility iff $q = 0$. If $p = 1/2$, then $d\mathcal{E}_p(q)/dq = 0$, and all values of q have the same expected utility. Thus $\mathcal{E}_p(p) \geq \mathcal{E}_p(q)$, for all $q \in [0,1]$, only if $p = 0$, $1/2$, or 1.

This means that if you were ever to find yourself giving a hypothesis a probability other than 0, $1/2$, or 1, then you could increase your expected cognitive utility by shifting your probability for that hypothesis to the nearest extreme value (either 0 or 1, as the case may be). But in fact we think that shifting probabilities to extreme values like this, without any evidence to support the shift, could *not* increase expected cognitive utility.

The possible responses to this problem are the same as the responses to the problem about utility functions that make evidence gathering irrational; and they suffer from the same defects. Thus one might say that scientists do not in fact have utility functions like the one Horwich proposed; that in fact their utility functions are such that, in the absence of new evidence, expected utility is always maximized by holding to their current probability function. (Such utility functions have been discussed in the Bayesian literature under the rubric of 'proper scoring rules' [Savage 1971].) This response makes it merely a contingent fact, which might have been otherwise, that arbitrary shifts in probability to extreme values do not increase expected cognitive utility. But this truth is surely not something that might have been otherwise.

The other possible response is to say that a utility function does not count as representing the cognitive goals of *science* unless it is a proper scoring rule. This succeeds in making it a necessary truth that arbitrary shifts in probability do not increase expected cognitive utility; but like the similar move used to make SVET necessary, success is here bought at the price of invoking a completely ad hoc assumption. The lack of any prior plausibility for the present assumption is indicated by the fact that it deems unscientific the utility function that Horwich proposed as representing scientific goals!

179

When cognitive consequences are taken to result from acceptance decisions, rather than from probabilities, no such ad hocery is needed. In an arbitrary shift in probability, one learns nothing about the states, in the sense of learning defined in Section 5.1.5. Hence the theorem of Section 5.1.6 shows that the expected utility of making an arbitrary shift cannot be higher than that of leaving current probabilities unchanged. By taking the subsequent actions to be decisions about what to accept, this theorem shows that arbitrary shifts also cannot increase expected *cognitive* utility. And this holds no matter what cognitive utility function a person might have.

To be fair to the probabilist explanation, I should point out that the utility functions that violate SVET are not proper scoring rules.[7] Thus if one counts only proper scoring rules as representing scientific values, no further assumption about the form of a cognitive utility function is needed to derive SVET. But this does not negate the fact that the probabilist must restrict cognitive utility functions in an ad hoc way, both to explain SVET and also to prevent arbitrary shifts in probability being rational.

To sum up this section: There are two attempts to explain SVET without invoking acceptance. The first, Good's, makes the mistake of conflating the cognitive goals of science with pragmatic goals. The second, the probabilist approach of Rosenkrantz and Horwich, postulates cognitive utilities determined by a scientist's utility function; this postulate, unlike the corresponding postulate concerning acceptance, does no work elsewhere in the philosophy of science. Furthermore, the probabilist approach requires additional ad hoc assumptions about the form of a scientific cognitive utility function. By contrast, the acceptance-based approach gets to explain SVET for free – no further assumptions need to be invoked.[8]

---

[7]Proof. Think of adopting a new probability function as an act, so that utilities depend on the act and state of nature, in the usual way. Then if the utility function is a proper scoring rule, the "act" chosen will be one that maximizes expected utility at the time it is chosen. (For a proper scoring rule is a utility function for probabilities that always makes the current probability function maximize expected utility.) We can now prove, in the usual way, that gathering evidence cannot reduce expected utility.

[8]For an argument that neither Popper nor Kuhn is able to explain SVET, see (Maher 1990c).

180

7.4 SUMMARY

I have given three reasons why acceptance is an important concept for the philosophy of science. Acceptance is needed to explain the history of science; it is needed to explain the role of alternative hypotheses in science; and it figures in the best account of how gathering evidence contributes to scientific goals.

8

Representation theorem

In Chapters 6 and 7, I assumed that rational scientists have utilities for cognitive consequences, as well as probabilities for scientific hypotheses. It is now time to defend that assumption. According to the preference interpretation of probability and utility given in Section 1.3, the assumption would be true if rational scientists have preferences that maximize expected cognitive utility, relative to some probability and cognitive utility functions. My argument will therefore consist in showing that the cognitive preferences of rational scientists can be so represented. This will be done by stating a suitable representation theorem.

It turns out that existing representation theorems are not quite suitable for this purpose. I will show why Savage's representation theorem is not suitable and then state a new representation theorem that does what is wanted. This representation theorem comes in two parts, the first (Section 8.2) establishing a representation for simple cognitive acts, and the second (Section 8.3) extending this to cognitive acts in general. I will describe the assumptions of the theorem in some detail, but proofs are relegated to appendixes.

Even without proofs, this chapter is technical. Readers willing to grant the possibility of a representation (even if only for the sake of argument) could skip this chapter and proceed directly to the discussion of scientific values in Chapter 9.

8.1 SAVAGE'S UNINTERPRETABLE ACTS

In Chapter 1, I mentioned Savage's (1954) representation theorem and discussed some of its postulates. Here I will show that Savage's theorem has a defect that makes it unsatisfactory to use as a basis for deriving cognitive utilities.

182

I begin with Savage's conception of an *act*. In Savage's theory, acts, states, and consequences are conceptualized in such a way that the act chosen, together with the state that obtains, uniquely determines what consequence will be obtained. Hence it is possible to think of acts as functions mapping the set of states onto the set of consequences. But Savage goes a step further, and *identifies* the set of acts with the set of functions from the states to the consequences. Thus every function from the states to the consequences is regarded as an act. I will call these functions *Savage acts*.

In many decision problems, some of these Savage acts do not correspond to any conceivable decision. In particular, this is true of cognitive decision problems. For example, let A be a proposition, and let $\langle A, t \rangle$ be the cognitive consequence that we can describe as "accepting A when A is true." Let f be the function that maps every state onto $\langle A, t \rangle$; that is, $f(x) = \langle A, t \rangle$, for all states x. Then f is a Savage act. But if A is not a necessary truth, there are states x in which A is false; and there is no conceivable decision that would give the consequence $\langle A, t \rangle$ in those states. Thus the Savage act f does not correspond to any conceivable decision. Savage acts like this, which correspond to no conceivable decision, I will call *uninterpretable* acts.

The Savage act f in this example is an instance of what is called a *constant act*, that is, an act whose value is the same for all states. In cognitive decision problems, it would appear that constant acts are typically uninterpretable. However, there are exceptions. It is reasonable to take the cognitive consequence of accepting the tautologous proposition as being the same whatever state obtains, since this proposition is true in all of them. Similarly, we can plausibly take the cognitive consequence of accepting the contradictory proposition to be the same whatever state obtains, since this proposition is false in every state. (Moreover, it captures no more of the truth about any one state than about any other.) Thus the decisions to accept the tautologous proposition and to accept the contradictory proposition are both interpretable constant acts.

Not all uninterpretable acts are constant acts. If $f(x) = \langle A, t \rangle$, for some state x in which A is false, then f is an

183

uninterpretable act, regardless of its values for other states. But all these functions are regarded as acts by Savage.

It makes no sense to speak of preferences regarding uninterpretable acts. For the notion of preference we are concerned with is a preference for *performing* one act rather than another, not a preference for (say) contemplating the mathematical form of one act rather than another; and the notion of performing an uninterpretable act is nonsense. This means that by including uninterpretable acts in the set of acts, Savage makes it logically impossible for preferences to be connected on the set of acts. Thus Savage's first postulate, which asserts that preferences are connected (Section 1.3), logically cannot be satisfied on the set of Savage acts.

In Section 1.5, I did allow that the connectedness axiom does not need to be a requirement of rationality, in order for representation theorems assuming this axiom to be useful. I said that if your preferences agree so far as they go with some connected preference rankings that satisfy the other postulates, then the representation theorem tells us that these connected preference orderings are representable by a p-u pair. Hence the representation theorem can be interpreted as telling us that rational preferences are represented by at least one p-u pair. But in saying that preferences satisfying the other axioms agree with some connected preference ordering, I was of course assuming that the acts were the sorts of things that it is possible to have preferences about; and that is not so when the class of acts includes uninterpretable acts.

Still, it remains the case that if your preferences satisfy the other postulates, there is a binary relation $\precsim$ that is connected on the set of all acts, satisfies the other postulates, and agrees with your preferences so far as they go. Savage's representation theorem shows that there is a probability function p and a utility function u such that $f \precsim g$ iff $EU(f) \leq EU(g)$, for all Savage acts f and g (where EU is expected utility calculated using p and u). We could then say that p and u represent your actual preferences, even though the relation $\precsim$ is not a possible preference relation.

I think this approach does succeed in salvaging something from Savage's representation theorem. But interpreted in this

184

way, Savage's representation theorem only establishes the *existence* of some p-u pair that represents rational preferences; it does not establish any uniqueness result. Even if your preferences were defined for all interpretable acts, Savage's representation theorem does not allow us to infer that your probability and utility functions are unique;[1] for Savage gets this uniqueness result by assuming connectedness on the set of Savage acts, and we have seen that it is logically impossible for your preferences to be connected on this set.[2]

The inclusion of uninterpretable acts also creates problems for some of Savage's other postulates. For example, the independence postulate asserts that if f, f', g, and g' satisfy certain conditions, and if $f \precsim g$, then $f' \precsim g'$. However, it might be that f and g are interpretable, while f' and/or g' are not; and in this case, it would be impossible to satisfy independence. We might get around this by saying that independence is only a requirement of rationality when all the acts involved are interpretable; but still, this is an inelegant way to construct a theory. One would rather have a representation theorem that does not require the assumption of preferences regarding uninterpretable acts.

8.2 SIMPLE COGNITIVE EXPECTED UTILITY

I will now state a representation theorem that is broadly similar to Savage's, but I will avoid Savage's reliance on uninterpretable acts. This representation theorem will show that rational preferences regarding simple cognitive acts correspond to maximizing expected utility relative to some probability and cognitive utility functions. In Section 8.3, the restriction to simple acts will be removed.

[1] A "unique" utility function here means one that is unique up to a positive affine transformation. This notion will be explained in Section 8.2.10.

[2] For Savage's use of the assumption that the preference relation extends to uninterpretable acts, see in particular the proof of the first theorem in Savage (1954, ch. 5). This proof requires that for each consequence a and for any state x there be an act f (in the domain of the preference relation) such that $f(x) = a$; and there will not generally be an interpretable act f satisfying this condition. Also, Savage's proof of the existence of a utility function (1954, ch. 5, sec. 3) depends on the assumption that for each consequence a there is a constant act that gives a in every state; and as we have seen, constant acts are often uninterpretable.

185

8.2.1 Notation

The set of *states* will be denoted by X. These are to be formulated in such a way that you are sure exactly one of them obtains. They are also to be sufficiently specific that they determine what consequence will result from each act under consideration. So, for example, if we are concerned with the act of accepting A, and the cognitive consequence of this act depends on the truth value of A, then the states must specify the truth value of A. If what is of cognitive value in accepting A depends on more than the truth value of A (e.g., because it matters how "close to the truth" A is when false), then the states need to specify more than the truth value of A. We saw an example of how to do this in Section 6.3.2, where the acts were to accept various hypotheses about the value of a parameter, and the states specified the true value of the parameter.

Events (or propositions) are taken to be sets of states, that is, subsets of X. The set of all events will be denoted by $\mathbf{X}$. (Here I am using the convention according to which a boldface letter denotes a set of subsets of the set denoted by the corresponding italic letter.) I do not assume that $\mathbf{X}$ contains every subset of X; I merely assume that $\mathbf{X}$ is a σ-algebra of subsets of X.[3]

The set of consequences will be denoted by Y. As always, the consequences need to specify everything of value in the situation. In cognitive decision problems, if all we cared about was whether the corpus we accept is true or false, then the consequences could be taken to be of the form "Accepting A when it is true" and "Accepting A when it is false." But if the value of accepting a false corpus varies depending on which state obtains (as it will if we care about "distance from truth"), then the consequence of accepting A when it is false should rather be taken to be of the form "Accepting A when state x obtains."

I will use $\mathbf{Y}$ to denote a σ-algebra of subsets of Y that contains all the singleton sets of Y. That is, for all $a \in Y$, $\{a\} \in \mathbf{Y}$.

The set of possible acts, or decisions, will be denoted by $\mathbf{D}$. In cognitive decision problems, this set will include acts of

---

[3]This means that $\mathbf{X}$ contains the empty set $\emptyset$, the set X, and is closed under complementation and countable unions. To be closed under complementation means that if $A \in \mathbf{X}$, then $\bar{A} \in \mathbf{X}$. To be closed under countable unions means that if $A_1, A_2, \ldots \in \mathbf{X}$, then $\bigcup_{i=1}^{\infty} A_i \in \mathbf{X}$.

accepting various potential corpora. As in Savage's theory, the elements of **D** will be taken to be functions from X to Y. However, unlike Savage's theory, I do not assume that **D** contains *every* function from X to Y. Exactly what **D** must include will be explained shortly.

A function f from X to Y is said to be *measurable* if, for all $B \in \mathbf{Y}$, $f^{-1}(B) \in \mathbf{X}$.[4] The acts in **D** will be required to be measurable. Thus if there are functions from X to Y that are not measurable, these are excluded from **D**. This is a merely formal requirement; it does not force any specific function to be excluded from **D**, since the choice of **X** and **Y** is up to us.

I turn now to a description, and discussion, of the various assumptions on which my first representation theorem rests.

8.2.2 Connectedness

The representation theorem will assume that the weak preference relation is connected on **D**. Formally, this is

Axiom 1. *For all $f, g \in \mathbf{D}$, either $f \precsim g$ or $g \precsim f$ (or both).*

The status of this assumption was explained in Section 1.5; it is not a requirement of rationality, but its assumption in a representation theorem is harmless, provided we do not interpret a representation theorem as showing that a rational person must have *unique* probability and utility functions.[5]

8.2.3 Two constant acts

Unlike Savage's representation theorem, the theorem to be stated here does not assume that for every $a \in Y$ there is a corresponding constant act in **D**. However, it is assumed that **D** contains (at least) two constant acts, one of which is strictly preferred to the other. As I noted in Section 8.1, this is a plausible assumption when we are dealing with cognitive decision problems. For there are always the options of accepting X (the tautologous proposition) and $\emptyset$ (the impossible proposition). In accepting X one accepts truth no matter what state obtains, and in accepting $\emptyset$ one accepts falsehood no matter what state

---

[4] $f^{-1}(B)$ is the set of all $x \in X$ such that $f(x) \in B$.
[5] Footnote 1 applies here also.

187

obtains. Furthermore, I imagine that most people would strictly prefer accepting necessary truth to accepting necessary falsehood.

The preference relation $\precsim$ can be extended in a natural way to the consequences that are values of constant acts. Thus if $a, b \in Y$, I shall interpret '$a \precsim b$' as meaning that there exist $f, g \in D$ such that $f \precsim g$, and for all $x \in X$, $f(x) = a$ and $g(x) = b$. Similarly for '$a \prec b$'. In this notation, the assumption that there are two constant acts, one strictly preferred to the other, can be expressed as

Axiom 2. *There exist $a, b \in Y$ such that $a \prec b$.*

8.2.4 Closure condition

If f and g are acts, then the following is also a conceivable act: First determine whether or not A is true, then choose f if A is true, and g otherwise. This latter act gives the same consequences as f for states in A, and gives the same consequences as g for states in $\bar{A}$. I will assume that for any $f, g \in D$ and for any $A \in \mathbf{X}$, there is always such an act. Formally:

Axiom 3. *For all $f, g \in D$ and $A \in \mathbf{X}$, there exists $h \in D$ such that $h = f$ on A, and $h = g$ on $\bar{A}$.*

We could call h a *mixture* of f and g, and then Axiom 3 can be expressed by saying that D must be closed under the mixing operation.

To see the import of this Axiom for cognitive decision problems, suppose that f is the act of accepting some hypothesis F, and g is the act of accepting a hypothesis G. Then h is the act of learning whether A is true and accepting F if it is, G if it is not. This act h will generally not be identifiable with the act of accepting any particular hypothesis, and thus Axiom 3 entails that in cognitive decision problems the class D cannot be limited to acts of acceptance.

The act h can be described as an *experiment*, since it consists of observing (or otherwise determining) whether or not a certain event obtains, and then making an inference conditional on that observation. Sometimes, of course, the mere making of

188

an observation is called an experiment; but the design of an experiment properly includes a specification of the inference to be drawn from each possible outcome, and so it is not unnatural to think of an experiment as consisting not just of an observation, but as including also a rule determining the inference to be drawn from the observation.

In general, I will call an act an experiment if it involves learning which element in some partition[6] $\{A_\lambda : \lambda \in \Lambda\}$ obtains (where Λ is an arbitrary index set), and for each $\lambda \in \Lambda$ there is a proposition B_λ that will be accepted if A_λ obtains. It is clear that the class of experiments is closed under mixing. That is to say, if f and g are two experiments, and h is the act that equals f on A and g on $\bar{A}$, then h is also an experiment. Thus in cognitive decision problems, Axiom 3 can be satisfied by taking $\mathbf{D}$ to be a class of experiments. Furthermore, the act of accepting a hypothesis B is a degenerate case of an experiment; it consists of learning the true element of the trivial partition $\{X\}$, and accepting B if X obtains (as it must). Consequently, Axiom 2 can also be satisfied by taking $\mathbf{D}$ to be a class of experiments. The axioms yet to be stated can likewise be satisfied when $\mathbf{D}$ is a class of experiments.

It is not necessary for $\mathbf{D}$ to contain *all* experiments in order for the assumptions of the representation theorem to be satisfied. In particular, the assumptions that have been stated thus far are satisfied by taking $\mathbf{D}$ to be the class of those experiments that have the following form: It is observed whether or not some event $A \in \mathbf{X}$ obtains, with X being accepted if A does obtain and $\emptyset$ being accepted otherwise. Of course, we would normally want to consider more interesting experiments than this, and my point here is merely to indicate the smallest class of experiments that satisfies the assumptions of the representation theorem.

Not all experiments are practical possibilities. For example, an experiment that involved observing whether or not all electrons (in all of space-time) have the same charge is not an experiment that is available to finite beings. However, even experiments like

---

[6] A partition is a set of mutually exclusive and jointly exhaustive events (or propositions). In addition, I will add the stipulation that a set of events counts as a partition only if each event is an element of $\mathbf{X}$.

this are conceptual possibilities, and often people will have preferences about them. If A is the proposition that all electrons have the same charge, then few would disagree that the experiment of observing whether A is true, and accepting A if it is true and $\bar{A}$ otherwise, is preferable to the experiment that makes the same observation but draws the opposite conclusion. This preference holds despite the fact that both of the experiments here compared are unrealizable in practice.

8.2.5 Transitivity

The representation theorem also assumes that the preference relation $\precsim$ is transitive. This is

Axiom 4. *For all* $f, g, h \in \mathbf{D}$, *if* $f \precsim g \precsim h$, *then* $f \precsim h$.

The status of this principle was discussed in Chapter 2. I do take it to be a requirement of rationality, though I conceded that this position is not entirely uncontroversial. I suggested that the most fruitful approach to resolving the controversy would be to see what consequences flow from the principle and its alternatives. The present work is drawing out some consequences of transitivity together with other principles.

8.2.6 Independence

My representation theorem also assumes the independence axiom:

Axiom 5. *For all* $f, f', g, g' \in \mathbf{D}$ *and* $A \in \mathbf{X}$, *if* $f = f'$ *on* A, $g = g'$ *on* A, $f = g$ *on* $\bar{A}$, $f' = g'$ *on* $\bar{A}$, *and* $f \precsim g$, *then* $f' \precsim g'$.

The status of this principle was discussed in Chapter 3. I showed there that independence follows from the intuitively attractive principle of synchronic separability. While this will not end all debate, my suggestion here too was that further progress toward consensus is most fruitfully sought by developing applications of the principle and its alternatives.

8.2.7 Generalized weak dominance

At this point, it is convenient to define a notion of conditional preference, as follows:

190

Definition 8.1. $f \precsim g$ given A *iff for all* $f', g' \in \mathbf{D}$, *if* $f' = f$ *on* A, $g' = g$ *on* A, *and* $f' = g'$ *on* $\bar{A}$, *then* $f' \precsim g'$.

Axioms 1 and 5 imply that for all $f, g \in \mathbf{D}$ and $A \in \mathbf{X}$, either $f \precsim g$ given A, or $g \precsim f$ given A (or both). If $f \precsim g$ given A, but $g \not\precsim f$ given A, then we say $f \prec g$ given A. Similarly, if $f \precsim g$ given A and $g \precsim f$ given A, then we say $f \sim g$ given A.

An event $A \in \mathbf{X}$ will be said to be a *null event* if the person is indifferent between all acts that differ only on A. The class of null events will be denoted by $\mathbf{N}$. The formal definition of $\mathbf{N}$ is

Definition 8.2. $A \in \mathbf{N}$ *iff* $A \in \mathbf{X}$ *and for all* $f, g \in \mathbf{D}$, $f \precsim g$ *given* A.

This is no more than a definition; it is not assumed that there actually are any null events. But null events, if there are any, will be assigned probability 0 by the representation theorem.

Given these two definitions, I can now state what I will call the principle of weak dominance, as follows:[7]

Weak dominance. If $f, g \in \mathbf{D}$, $a, b \in Y$, $f = a$, and $g = b$, then for all $A \in \mathbf{X} \setminus \mathbf{N}$, $f \precsim g$ given A iff $a \precsim b$.

I think that weak dominance is an uncontroversial principle of rationality. But its restriction to consequences that are the values of constant acts seems unnecessarily restrictive. To see how a suitable generalization may be effected, note that the following is equivalent to the weak dominance principle:

If $f, g \in \mathbf{D}$, $a, b \in Y$, $f = a$, and $g = b$, then for all $A, B \in \mathbf{X} \setminus \mathbf{N}$, $f \precsim g$ given A iff $f \precsim g$ given B.

From this formulation, it is a natural step to the following generalization, which is not restricted to consequences that are the values of constant acts:

Axiom 6. *For all* $f, g \in \mathbf{D}$, $a, b \in Y$, *and* $A, B \in \mathbf{X} \setminus \mathbf{N}$, *if* $f = a$ *on* A, $f' = b$ *on* A, $g = a$ *on* B, *and* $g' = b$ *on* B, *then* $f \precsim f'$ *given* A *iff* $g \precsim g'$ *given* B.

---

[7] The notation $\mathbf{X} \setminus \mathbf{N}$ denotes the set of elements of $\mathbf{X}$ that are not in $\mathbf{N}$. In words, it is the set of nonnull events.

191

If it is assumed that every consequence is the value of a constant act, then Axiom 6 is in fact equivalent to weak dominance. So it does indeed capture the essence of weak dominance, without the restriction to consequences that are the values of constant acts. It seems to be no more controversial than weak dominance itself, and the representation theorem will assume it.

8.2.8   Qualitative probability

We can say that event B is more probable for you than event A, just in case you prefer the option of getting a desirable prize if B obtains, to the option of getting the same prize if A obtains. But for this "more probable than" relation to be well defined, it must be the case that whether you prefer a prize to be conditional on A or B does not depend on what the prize is, just so long as the prize is desirable. This appears a reasonable condition; and I will now state an axiom that captures it formally.

The option of getting a desirable prize if A obtains can be regarded as an act f such that $f = b$ on A, $f = a$ on $\bar{A}$, and $a \prec b$. Similarly, the option of getting the same desirable prize if B obtains is an act g such that $g = b$ on B and $g = a$ on $\bar{B}$. You prefer the prize to be conditional on B iff $f \prec g$. We could then say that your preference is independent of what the prize is, so long as substituting c for a and d for b does not reverse your preference, for all c and d such that $c \prec d$.

While this is a reasonable condition, the formulation just given is unnecessarily restrictive, since it applies only when the prizes are the values of constant acts. (This is due to the conditions that $a \prec b$ and $c \prec d$.) It should not make any difference if there are states in $\bar{A} \cap \bar{B}$ in which no act gives b, or states in $A \cap B$ for which no act gives a. Similarly for c and d. To see how to remove this gratuitous restriction, note that if f and g are as in the preceding paragraph, we can replace the requirement that $a \prec b$ by the requirement that $f \prec g$ given $A \setminus B$. The latter requirement is equivalent to the former if a and b are the values of constant acts,[8] but the latter requirement can hold in other cases as well. So the condition stated in the

---

[8] Assuming $A \setminus B \notin \mathbf{N}$. There is no need to discuss here the case when $A \setminus B \in \mathbf{N}$, because Axiom 7 says nothing about that case. If $A \setminus B \in \mathbf{N}$, then $f \sim g$ given $A \setminus B$, so the antecedent of Axiom 7 is false.

192

preceding paragraph can be generalized in the following way:

Axiom 7. *For all $f, f', g, g' \in \mathbf{D}$; $a, b, c, d \in Y$; and $A, B \in \mathbf{X}$: if*

- *$f = b$ on A and a on $\bar{A}$,*
- *$g = b$ on B and a on $\bar{B}$,*
- *$f' = d$ on A and c on $\bar{A}$,*
- *$g' = d$ on B and c on $\bar{B}$,*
- *$f \prec g$ given $A \setminus B$,*
- *$f' \prec g'$ given $A \setminus B$, and*
- *$f \underset{\sim}{\precsim} g$,*

then $f' \underset{\sim}{\precsim} g'$.

The generalization does not affect the intuitive rationale for this condition, namely: Whether you would rather have a desirable prize riding on A or B should not depend on what the prize is, just so long as it is desirable.

8.2.9 Continuity

For the next axiom, it seems best to first state it, then give illustrations and discuss its import.

Axiom 8. *For all $f, g \in \mathbf{D}$, $a \in Y$, and $A \in \mathbf{X}$, if $f \prec g$ then there exists in $\mathbf{X}$ a partition $\{A_1, \ldots, A_n\}$ of A such that for $i = 1, \ldots, n$:*

- *If $f_i = a$ on A_i and f on $\bar{A}_i$, then $f_i \prec g$.*
- *If $g_i = a$ on A_i and g on $\bar{A}_i$, then $f \prec g_i$.*

Here is an example of the application of Axiom 8 in a cognitive decision problem. It makes use of these propositions from Section 6.3.1:

H_C: The electrostatic force falls off as the nth power of the distance, for some n between 1.98 and 2.02.

H'_C: The electrostatic force falls off as the nth power of the distance, for some n between 1.9 and 2.1.

Let the f and g of Axiom 8 be the acts of accepting H'_C and H_C, respectively. Also let A be the proposition that the electrostatic force falls off exactly as the second power of the distance, and let a be the cognitive consequence of accepting A when it

193

is true. We know that for Cavendish, after he had conducted his experiment, $f \prec g$. Axiom 8 thus requires that there be a partition $\{A_1, \ldots, A_n\}$ of A, such that for $i = 1, \ldots, n$,

If $f_i = a$ on A_i and f on $\bar{A}_i$, then $f_i \prec g$.

Now consider the trivial partition of A that consists of just A itself; that is, $A_1 = A$. Then f_1 is the act of accepting A if A is true, and otherwise accepting H'_C.[9] It may be that for Cavendish, $f_1 \prec g$. Intuitively, this would occur if Cavendish thought A very improbable; for then he would think f_1 is most likely to give whatever consequence f would give, and he has $f \prec g$. If this possibility is realized, the first clause of Axiom 8 is satisfied for this example.

But perhaps Cavendish has $g \precsim f_1$. If so, we will consider a nontrivial partition of A. For example, Cavendish might know that a coin is to be tossed ten times. Then we can let A_1 be the proposition that A is true and that all ten tosses land heads, let A_2 be the proposition that A is true and only the first nine tosses land heads, and so on. In this way, we get a partition of A with 2^{10} members. Now f_1 is the act that gives the consequence a if A is true and all ten tosses of the coin land heads; otherwise f_1 gives whatever consequence f would give. Similarly for $f_2, \ldots, f_{2^{10}}$. Since each partition element is very unlikely, each f_i is overwhelmingly likely to give the same consequence as f would give; and so we might expect to have $f_i \prec g$, for all $i = 1, \ldots, 2^{10}$. If so, the first clause of Axiom 8 is satisfied. If not, try an even finer partition of A.

Thus we see the import of Axiom 8: It is likely to hold only if the set of states is sufficiently numerous that propositions can be subdivided into arbitrarily fine divisions. That is why this axiom is dubbed "continuity." This should not be a controversial assumption, because we can always enrich the class of states if necessary. Of course, a person might not have preferences over all the options definable in the enriched set of states, but we are not insisting that a person's preferences be connected.

Axiom 8 also requires that there be no consequence so desirable that adding even the tiniest chance of getting it is enough

[9]This is an experiment, in the sense of Section 8.2.4.

194

to reverse preferences between acts. This amounts to saying that the decision maker does not view any consequence the way Pascal, in his famous wager, viewed the prospect of getting to heaven. I think this is plausible in cognitive contexts, at least. The best possible cognitive consequence is presumably learning everything one would like to know, and few would be inclined to reverse cognitive preferences for an infinitesimal chance of attaining this consequence.

So far I have illustrated and discussed only the first clause of Axiom 8, but the second clause is similar. With a as above, the g_i defined in this clause would actually be more attractive options than g itself, and so it is unproblematic that $f \prec g_i$. A more interesting case arises when a is an undesirable consequence. For example, suppose a is the consequence of accepting the contradictory proposition $\emptyset$. Using the partition $\{A_1, \ldots, A_{2^{10}}\}$ defined three paragraphs back, each g_i has only a tiny chance of resulting in the acceptance of a contradiction; it is almost certain that g_i will give the same consequence as g would. And so we might well expect that $f \prec g_i$, for all $i = 1, \ldots, 2^{10}$. If not, try a finer partition.

Here we see that the second clause of Axiom 8 requires that there be no consequence so undesirable that adding even the tiniest chance of getting it is enough to reverse preferences between acts. This is, I think, plausible in cognitive contexts. Perhaps the worst cognitive consequence is to accept a contradiction, and I think most of us can accept a sufficiently small risk of doing that.

8.2.10 Representation theorem

A representation theorem for expected utility has both an *existence* and a *uniqueness* component. In a typical representation theorem, such as Savage's, the existence component says that there is a probability function p and a utility function u such that for all $f, g \in \mathbf{D}$,

$$f \precsim g \quad \text{iff} \quad EU(f) \leq EU(g),$$

where the expected utility EU is calculated using p and u. This establishes that there is a representation. The uniqueness component typically says that the probability function p is unique;

that is, p cannot be replaced by any other function without destroying the representation. The uniqueness component also typically says that the utility function u is unique in the sense that if u' can replace u without destroying the representation, then there exist constants ρ and σ, with $\rho > 0$, such that $u' = \rho u + \sigma$.[10] This relation between u and u' is called a *positive affine transformation*, and thus the uniqueness result is also expressed by saying that u is unique up to a positive affine transformation.

The representation theorem that can be proved on the basis of Axioms 1 to 8 differs from this paradigm of a representation theorem in both its existence and uniqueness components. These differences will now be explained.

An act is said to be *simple* if it has only finitely many possible consequences. That is to say, f is simple if there are only finitely many $a \in Y$ such that, for some $x \in X$, $f(x) = a$. If we suppose that the cognitive consequence of accepting a hypothesis depends only on the truth value of the hypothesis, then acceptance of a hypothesis has only two possible consequences and hence is a simple act. Under the same supposition, experiments are also simple acts, provided the partition whose true element will be ascertained is finite. The representation theorem to be stated here provides an expected utility representation for simple acts only. Thus the existence claim of this representation theorem is weaker than the paradigm described above.[11]

Let Y^* be the set of all $a \in Y$ such that, for some $f \in D$, $f^{-1}(a) \notin N$. Thus Y^* is the set of all consequences that can be obtained with positive probability. So the cognitive consequence "accepting A when A is true" is a member of Y^* only if $A \notin N$. The uniqueness part of the representation theorem to be stated here says that if p and u are probability and utility functions that give the expected utility representation (on simple acts),

---

[10] More fully: $u'(a) = \rho u(a) + \sigma$, for all $a \in Y$.

[11] Axioms 1–8 do not entail an expected utility representation for all acts. Savage (1954, p. 78) gives an example in which Axioms 1–8 are all satisfied but no complete expected utility representation exists. Savage's example was one in which the probability p was not countably additive, and Savage speculated that this failure of countable additivity was an essential feature of his example. However, Seidenfeld and Schervish (1983, p. 404) give an example in which p is countably additive, and Axioms 1–8 are satisfied, but still there is no expected utility representation of the preferences.

and if u' is another utility function that can be substituted for u while preserving that representation, then there exist constants ρ and σ, $\rho > 0$, such that $u' = \rho u + \sigma$ on Y^*.[12] This can be described by saying that u is unique, up to a positive affine transformation, on Y^*. The limitation to Y^* makes this uniqueness result weaker than the paradigm described in the first paragraph of this section.

With these explanations, I can now state the theorem.

Theorem 8.1. *If Axioms 1–8 hold, then there exists a probability function p on $\mathbf{X}$, and a utility function u on Y, such that for all simple $f, g \in \mathbf{D}$,*

$$f \precsim g \quad iff \quad EU(f) \leq EU(g),$$

where EU is expected utility calculated using p and u. The probability function is unique, and the utility function is unique, up to a positive affine transformation, on Y^*.

The proof of this theorem is given in Appendix B.

If the acceptance of a hypothesis were a simple act, then Theorem 8.1 would provide a satisfactory foundation for the decision-theoretic account of rational acceptance.

8.3 GENERAL COGNITIVE EXPECTED UTILITY

In Section 6.3, I noted that the cognitive utility of accepting a false hypothesis is often held to depend on how close the hypothesis is to the truth. As the example discussed in that section illustrated, this view entails that accepting a hypothesis may have infinitely many possible consequences. If that is so, then accepting a hypothesis is not a simple act. In that case, Theorem 8.1 does not establish an expected utility representation for preferences regarding acts of acceptance.

However, with some additional assumptions, Theorem 8.1 can be strengthened so that it does provide an expected utility representation for acts of acceptance, even when those acts are not assumed to be simple. In this section, I will state the additional assumptions that are needed, and then the stronger representation theorem that they allow to be proved.

[12] More fully: $u'(a) = \rho u(a) + \sigma$, for all $a \in Y^*$.

197

I am not here assuming any theory of what it means for a false hypothesis to be more or less close to the truth. That topic will be taken up in Chapter 9. The point of invoking the idea here is merely to make it plausible that acceptance need not be a simple act. The representation theorem shows that we can speak meaningfully of cognitive utilities under these circumstances, without having any prior theory of distance from truth. Consequently, in Chapter 9 I will be able to define the notion of distance from truth in terms of cognitive utility.

8.3.1 Dominance for countable partitions

The Axioms of Section 8.2 entail the following (cf. Theorem B.2):

If $\{A_1, \ldots, A_n\}$ is a partition of A, and if $f \precsim g$ given A_i for all $i = 1, \ldots, n$, then $f \precsim g$ given A.

This is a type of dominance condition. A somewhat weaker condition is obtained by replacing $\precsim$ with $\prec$ in the antecedent of the conditional. This gives

If $\{A_1, \ldots, A_n\}$ is a partition of A, and if $f \prec g$ given A_i for all $i = 1, \ldots, n$, then $f \precsim g$ given A.

I will assume that this weakened form of the dominance condition applies even when the partition is countably infinite. So I assume

Axiom 9. *For all $f, g \in \mathbf{D}$ and $A \in \mathbf{X}$, if $\{A_1, A_2, \ldots\}$ is a partition of A, and if $f \prec g$ given A_i for all i, then $f \precsim g$ given A.*

Axiom 9 is, I think, immensely attractive; so much so that an argument for it would seem to be redundant. However, this axiom, together with the preceding ones, entails that the probability function p (whose existence is guaranteed by Theorem 8.1) is countably additive. This means that if $A_1, A_2, \ldots$ is a sequence of disjoint (or mutually exclusive) events, then $p(\bigcup_{i=1}^{\infty} A_i) = \sum_{i=1}^{\infty} p(A_i)$. By contrast, Axioms 1–8 merely entail that p is finitely additive; that is, they guarantee that $p(\bigcup_{i=1}^{n} A_i) = \sum_{i=1}^{n} p(A_i)$ for finite n, but not the extension of this to infinite sets of events. The proof that Axiom 9 entails countable additivity is given in Appendix D.

198

In the mathematical theory of probability and in applications of it in the sciences, it is generally assumed that probability is countably additive. Nevertheless, de Finetti has argued that we should not take countable additivity of probability to be a condition of rationality. One of his reasons for this is that if one knows an integer will be selected randomly, one might want to say that each integer has the same probability of being chosen and that the probability of some integer or other being chosen is 1. This violates countable additivity. De Finetti's other reason for rejecting countable additivity as a rationality condition is that he thinks there is no compelling argument for it, other than mathematical convenience. I will address these two reasons in turn.

If you told me you had selected an integer "at random," I would think 5 more likely than 5 million. Of course, sometimes nature "chooses" an integer, such as the number of stars in the universe or the number of hairs on my head. But here again, I have no inclination to judge all integers as equally likely to be selected. In fact, I have not been able to think of an actual case where it seems natural to assign equal probabilities to all elements of a countable partition. Perhaps this is a failure of imagination on my part; but at the very least it does indicate that there are few situations in which it is natural to assign probabilities in this way.

Turning now to de Finetti's claim that there is no argument for countable additivity, other than convenience: I would rebut this by noting that Axiom 9, together with other conditions accepted by de Finetti, entails that probabilities should be countably additive.

There is also an ad hominem argument: De Finetti accepts Dutch book arguments, and we can give a Dutch book argument for countable additivity. For example, suppose an integer is to be selected, and as de Finetti wishes to allow, you give each integer probability zero of being selected but are sure some integer will be selected. If you post these probabilities in a de Finetti–style Dutch book setup (cf. Section 4.6.4), you will have to accept bets in which you lose $1 if the integer selected is n and gain nothing if it is not n. But accepting all such bets gives you a sure loss of $1. (We can also make the bets each have a positive expected utility for you and still ensure that together they produce a sure loss.) Perhaps it will be objected here that finite agents cannot make infinitely many bets. But the bets need not be made serially; and even if finitude did preclude making infinite sets of bets, this does not seem to be a relevant objection to a thought experiment concerned with rationality conditions. Thus de Finetti cannot consistently reject countable additivity.

8.3.2 Acts and consequences

Besides Axiom 9, there are two more axioms that I need in order to prove the representation theorem for general cognitive utility. These two axioms require **D** and Y to have a certain structure. I think that the best procedure here is for me to first present what I will call the *favored interpretation* of **D** and Y; later I will show that this interpretation of **D** and Y satisfies the axioms yet to be stated.

I begin with the favored interpretation of Y. On this interpretation, for each nonempty $A \in \mathbf{X}$, Y contains the cognitive consequence "accepting A when A is true." I will continue to use the notation $\langle A, t \rangle$ to refer to this consequence. Since consequences must be complete in all respects that are of value (Section 1.1), I am assuming here that if you accept A and it is true, the satisfaction of your cognitive goals is not dependent on which state in A is the true state. For example, if the only thing you accept is that all ravens are black, and if all ravens are indeed black, then on this assumption your cognitive utility is not dependent on whether all swans are white, or on how many ravens there are, or on anything else. (This assumption will be defended in Section 9.3.)

On the favored interpretation of Y it also includes, for each nonempty $A \in \mathbf{X}$ and $x \in \bar{A}$, the cognitive consequence "accepting A when the true state is x." This consequence will be denoted $\langle A, x \rangle$. In identifying the consequences of accepting a false hypothesis in this way, we make no assumptions about what factors influence the cognitive value of accepting a false hypothesis; we merely assume that the states are sufficiently specific to determine the values of whatever the relevant factors may be.

200

Acceptance of the empty set (or contradictory proposition) $\emptyset$ was made an exception in each of the preceding two paragraphs. That is because, on the one hand, $\emptyset$ is true in no state; and on the other hand, as argued in Section 8.1, accepting $\emptyset$ is a constant act; it gives the same consequence in every state. This consequence will be denoted $\langle \emptyset, f \rangle$.

On the favored interpretation, Y contains all the consequences just mentioned, and no others. Thus Y is

$$\{\langle A, t \rangle : A \in \mathbf{X} \setminus \{\emptyset\}\} \cup \{\langle A, x \rangle : A \in \mathbf{X} \setminus \{\emptyset\}, x \notin A\} \cup \{\langle \emptyset, f \rangle\}.$$

The criterion for identity of cognitive consequences will be just the usual criterion of identity for ordered pairs, namely

$$\langle \varphi_1, \psi_1 \rangle = \langle \varphi_2, \psi_2 \rangle \quad \text{iff} \quad \varphi_1 = \varphi_2 \text{ and } \psi_1 = \psi_2.$$

I turn now to the interpretation of $\mathbf{D}$. Let a *countable experiment* be an experiment in which the true element of some countable partition $A_1, A_2, \ldots$ is ascertained, and for each $i = 1, 2, \ldots$ there is a proposition B_i that will be accepted if A_i obtains. In the favored interpretation, $\mathbf{D}$ contains all countable experiments.

It may be that only finitely many of the A_i in a countable partition are nonempty; hence the class of countable experiments includes those experiments associated with a finite partition. In particular, if only one of the A_i is nonempty, the countable experiment is identical with the decision to simply accept some hypothesis. Thus acceptance decisions are also included in the class of countable experiments.

In addition to countable experiments, the favored interpretation of $\mathbf{D}$ has $\mathbf{D}$ containing a special experiment, which I will denote f_T. This experiment consists of ascertaining which state obtains, and then accepting that the true state does obtain. So the definition of f_T is that for all $x \in X$, $f_T(x) = \langle \{x\}, t \rangle$. Since Axiom 8 forces X to be uncountable,[13] f_T is not a countable experiment; it is an uncountable experiment.

Besides countable experiments and f_T, $\mathbf{D}$ must contain mixtures of these acts, as required by Axiom 3. The favored interpretation of $\mathbf{D}$ is that it includes all these acts, *and no others*.

[13] As clause (iv) of Theorem B.1 shows.

As a result of these stipulations, $\mathbf{D}$ is the class of all acts f that have the following form:

There is a countable partition $\{A_1, A_2, \ldots\}$ such that f involves ascertaining which event A_i obtains and also, in case A_1 obtains, ascertaining which state obtains. If state $x \in A_1$ obtains, then $\{x\}$ is accepted; and for each $i = 2, 3, \ldots$ there is a proposition B_i that will be accepted if A_i obtains.

It is to be understood here that only one of $A_1, A_2, \ldots$ need be nonempty.

I will draw on this favored interpretation of $\mathbf{D}$ and Y in discussing the reasonableness of the two remaining axioms.

8.3.3 Density of simple acts

The next axiom is easy to state.

Axiom 10. *For all $f, g \in \mathbf{D}$, if $f \prec g$, then there exists a simple act $h \in \mathbf{D}$ such that $f \precsim h \precsim g$.*

An alternative way of saying the same thing is to say that the simple acts are dense in $\mathbf{D}$.

If Axioms 1–8 are satisfied, then the following conditions are jointly sufficient (but not necessary) for Axiom 10:

(i) There are "best" and "worst" acts; that is, there exist $k, l \in \mathbf{D}$ such that, for all $f \in \mathbf{D}$, $k \precsim f \precsim l$.

(ii) For all $f \in \mathbf{D}$, if $k \prec f$ or $f \prec l$, then there exists a simple act $h \in \mathbf{D}$ such that $h \precsim f$ or $f \precsim h$ (respectively).

A proof of this is given in Appendix C. I will now show that both of these conditions can be expected to hold, given the favored interpretation of $\mathbf{D}$ and Y.

The act f_T gives true and complete information about the world,[14] and so it can be expected to be at least weakly preferred to all other experiments; that is, $f \precsim f_T$ for all $f \in \mathbf{D}$. At any rate, this will be so if the preferences reflect the scientific concerns for truth and informativeness. Hence there is a "best" act,

---

[14] At least, it is complete relative to the set X of states, which is all that matters here.

202

as (i) requires. Furthermore, the act of accepting the contradictory proposition $\emptyset$ is surely not preferable to any other cognitive act; so if $f_\emptyset$ is the act of accepting $\emptyset$, we have $f_\emptyset \precsim f$, for all $f \in \mathbf{D}$. So there is a "worst" act too, and hence (i) is satisfied.

Turning now to (ii), suppose that $f \prec f_T$. For any positive integer n and finite partition $A_1, \ldots, A_n$ there is a possible experiment g_n that consists of ascertaining which A_i obtains, and accepting the one that does obtain. It is plausible that as n is increased without bound, and provided the events A_i are chosen suitably, the g_n would approach f_T in the preference ranking; in that case, there will be some n such that $f \prec g_n$. Now the possible consequences of g_n are $\langle A_1, t \rangle, \ldots, \langle A_n, t \rangle$, and hence are finite; thus g_n is a simple act. Thus the part of (ii) dealing with the "best" act does hold. Furthermore, as I argued in Section 8.1, the act $f_\emptyset$ is a constant act; hence it is a simple act; so if $f_\emptyset \prec f$, $f_\emptyset$ itself is a simple act such that $h \precsim f$. Thus both parts of (ii) hold.

So on the favored interpretation of $\mathbf{D}$ and Y, both (i) and (ii) can be expected to hold; from which it follows that Axiom 10 holds.

8.3.4 Structural assumptions

The final axiom imposes four conditions on the structure of $\mathbf{D}$, Y, and $\mathbf{X}$. I will first explain how each condition is satisfied on the favored interpretation of $\mathbf{D}$ and Y, then bring these conditions together to state the axiom.

Acceptance of a proposition A is an act all of whose consequences are (on the favored interpretation) of the form $\langle A, \psi \rangle$, for some ψ. Since $\langle A, \psi \rangle \neq \langle B, \psi \rangle$ if $A \neq B$, it follows that the possible consequences of accepting A are disjoint from the possible consequences of accepting any other proposition B. Using '$f(X)$' to denote the set of possible consequences of act f,[15] the fact just established can be expressed by saying that if f and g are the acts of accepting two different hypotheses, then $f(X) \cap g(X) = \emptyset$.

On the favored interpretation, it is also true that each consequence in Y is a possible consequence of accepting some

[15]More fully: $f(X) = \{y : f(x) = y, \text{ for some } x \in X\}$.

proposition or other; for each element of Y has the form $\langle A, t \rangle$, $\langle A, x \rangle$, or $\langle \emptyset, f \rangle$, and the first two of these are possible consequences of accepting A, while $\langle \emptyset, f \rangle$ is the (inevitable) consequence of accepting $\emptyset$. Letting $\mathbf{E}$ be the class of all acts of accepting some proposition or other, we can then say that $\bigcup_{f \in \mathbf{E}} f(X) = Y$. Combining this result with that of the preceding paragraph, we have that

(i) $\{f(X) : f \in \mathbf{E}\}$ is a partition of Y.

Now let f be any element of $\mathbf{E}$. This means there is some proposition A such that f is the act of accepting A. Thus one possible consequence of f is $\langle A, t \rangle$ (or $\langle A, f \rangle$ if $A = \emptyset$), and all other possible consequences, if there are any, are of the form $\langle A, x \rangle$. Now a consequence of the form $\langle A, x \rangle$ can only be obtained if x is the actual state. So we have that for each $f \in \mathbf{E}$ there is at most one exception to the rule that all possible consequences of f are consequences that can be obtained in just one state. This result can be expressed formally by saying:

For all $f \in \mathbf{E}$, there is at most one $y \in Y$ such that $f^{-1}(y)$ contains more than one element.

The consequences of the act f_T are all of the form $\langle \{x\}, t \rangle$, and so each possible consequence of f_T is obtainable in just one state. So for all $y \in Y$, $f_T^{-1}(y)$ contains at most one element. Letting $\mathbf{E}'$ be the set whose sole member is f_T, we now have the following strengthening of the preceding result.

For all $f \in \mathbf{E} \cup \mathbf{E}'$, there is at most one $y \in Y$ such that $f^{-1}(y)$ contains more than one element.

So a fortiori we have

(ii) For all $f \in \mathbf{E} \cup \mathbf{E}'$, there are at most countably many $y \in Y$ such that $f^{-1}(y)$ contains more than one element.

In Section 8.3.2, I said that on the favored interpretation $\mathbf{D}$ is the closure under mixing of the set of all countable experiments, together with f_T. It is easy to show that on this interpretation, the following is true.

204

(iii) For all $f \in \mathbf{D}$, there is a sequence $f_1, f_2, \ldots$ of elements of $\mathbf{E} \cup \mathbf{E}'$, and a partition $A_1, A_2, \ldots$ of X, such that for all positive integers i, $f = f_i$ on A_i.

We have already observed that the possible consequences of f_T are all of the form $\langle \{x\}, t \rangle$. The only way any of these can be a consequence of accepting a hypothesis A is if $A = \{x\}$ for some $x \in X$; and in that case there is just one consequence that f_T and accepting A have in common. (For all other consequences of accepting A are of the form $\langle \{x\}, f \rangle$.) Thus for all $h \in \mathbf{E}$, $h(X) \cap f_T(X)$ contains at most one element, and it is of the form $\langle \{x\}, t \rangle$. Now for any act $f \in \mathbf{D}$, $f^{-1}\langle \{x\}, t \rangle$ is either $\{x\}$ or $\emptyset$. Consequently, we have

For all $f \in \mathbf{D}$ and $h \in \mathbf{E}$, $f^{-1}[h(X) \cap f_T(X)]$ contains at most one element.

Since $\mathbf{E}' = \{f_T\}$, we therefore have

(iv) $\mathbf{E}'$ contains at most one element, and for all $f \in \mathbf{D}$, $h \in \mathbf{E}$, and $h' \in \mathbf{E}'$, $f^{-1}[h(X) \cap h'(X)]$ contains at most one element.

Putting together these four results, we have that on the favored interpretation, this axiom is satisfied:

Axiom 11. *There exist* $\mathbf{E}, \mathbf{E}' \subset \mathbf{D}$ *such that*
(i) $\{h(X) : h \in \mathbf{E}\}$ *is a partition of* Y.
(ii) For all $h \in \mathbf{E} \cup \mathbf{E}'$, *there are at most countably many* $y \in Y$ *such that* $h^{-1}(y)$ *contains more than one element.*
(iii) For all $f \in \mathbf{D}$, *there is a sequence* $f_1, f_2, \ldots$ *of elements of* $\mathbf{E} \cup \mathbf{E}'$, *and a partition* $A_1, A_2, \ldots$ *of* X, *such that for all positive integers* i, $f = f_i$ *on* A_i.
(iv) $\mathbf{E}'$ *contains at most one element, and for all* $f \in \mathbf{D}$, $h \in \mathbf{E}$, *and* $h' \in \mathbf{E}'$, $f^{-1}[h(X) \cap h'(X)]$ *contains at most one element.*

It is possible to weaken Axiom 11 in various ways without affecting the representation theorem (other than to make the proof more complicated). For example, it would not matter if $\mathbf{E}'$ had any finite cardinality. Also, it is possible to weaken the requirement in clause (iv) that $f^{-1}[h(X) \cap h'(X)]$ contain at

205

most one element. Just how far such weakenings can go is an open question. However, I show in Section D.5 that Axiom 11 cannot be entirely eliminated.

8.3.5 Representation theorem

The expected utility representation whose existence is given by Theorem 8.1 was limited to preferences concerning simple acts. With the addition of Axioms 9–11, we obtain a representation theorem without that limitation; it applies to the preference relation on **D**, without restriction.

Also, the utility function whose existence is given by Theorem 8.1 was unique only up to a positive affine transformation on Y^* and was quite arbitrary on $Y \setminus Y^*$. The utility function whose existence can now be derived satisfies the following much stricter uniqueness condition: If p and u are probability and utility functions that give the expected utility representation of preference on **D**, then u' is another utility function that can be substituted for u while preserving that representation iff there exist real numbers ρ and σ, $\rho > 0$, such that for all $f \in \mathbf{D}$, the set of all x such that

$$u[f(x)] = \rho u'[f(x)] + \sigma$$

is an event with probability 1. I will describe this by saying that u is unique, up to a positive affine transformation, *almost everywhere on Y.*[16]

So the representation theorem is the following:

Theorem 8.2. *If Axioms 1–11 hold, then there exists a probability function p on $\mathbf{X}$, and a utility function u on Y, such that for all $f, g \in \mathbf{D}$,*

$$f \precsim g \quad \text{iff} \quad EU(f) \le EU(g),$$

where EU is expected utility calculated using p and u. The probability function is unique; and the utility function is unique, up to a positive affine transformation, almost everywhere on Y.

---

[16]There is also a technical requirement: The composition $u' \circ f$ of functions u' and f must be measurable; otherwise the expected utility of act f would not be defined.

206

The proof of Theorem 8.2 is given in Appendix D.

Thus we now have a representation theorem that establishes an expected utility representation for cognitive preferences even when the acceptance of a hypothesis is not regarded as a simple act. This provides a vindication of the decision theoretic account of rational acceptance that I gave in Section 6.3. It also provides the foundation for the following discussion of scientific values.

9

Scientific values

Chapter 6 put forward the idea that scientists are concerned to accept theories that are informative and close to the truth; and it was assumed there that these values can be represented by a cognitive utility function. Chapter 8 has provided support for the idea that cognitive values can be represented by a cognitive utility function. However, a person's preferences can satisfy all the axioms of Chapter 8 without that person having the sorts of values we would call scientific; in this case, the person's preferences are representable by a cognitive utility function, but that function does not reflect scientific values.

This chapter investigates the question of what sort of cognitive values count as scientific. I articulate and defend the view put forward in Chapter 6: that science values informativeness and closeness to the truth and nothing else.

9.1 TRUTH

The notion of truth figures centrally in my account of scientific values. This notion is sometimes regarded as a dubious one, in need of clarification before it can be used with a good conscience. So I should perhaps begin by putting such worries to rest. Fortunately, that is easily done.

All we need to know about truth is that it has what Blackburn (1984) calls the *transparency property*. This property is that statements of the form 'It is true that *p*' mean the same as '*p*' (where *p* is to be replaced by any sentence).[1] For example, 'It is true that sodium burns with a yellow flame' means the same

---

[1] It would be customary to restrict *p* here to sentences that do not themselves contain the predicate 'is true', or to impose some other restriction to the same effect. But I think it is more elegant to follow the proposal of Gupta (1989), which does not require such a restriction, and instead provides a method for understanding circular definitions.

as 'Sodium burns with a yellow flame'. Since the meaning of the latter sentence is unproblematic (if anything is), so is the meaning of the former. Consequently, talk of truth need not be mysterious or problematic.

In my experience, those who find truth a problematic notion are often failing to distinguish between truth and certainty. For example, it is sometimes said that since we cannot be certain of anything, therefore there is no such thing as absolute truth or falsity. If "absolute truth" means certainty, this little argument is sound, but trivial. If "absolute truth" means truth – that is, if it has the transparency property – then the inference is a non sequitur. In fact, if one accepts the logical principle of excluded middle

p or not-p

then it follows that every statement is either true or false.

Proof. Let p be any statement. From excluded middle and the transparency property we obtain

p is true or not-p is true.

Since 'p is false' means 'not-p is true', this is the desired result.

I mentioned that what I say about truth in this book requires only the transparency principle in order to be intelligible. For example, I hold that the scientific utility of accepting a statement A is at least as great when A is true as when A is false. This claim can be restated by saying that the scientific utility for 'A is accepted, and A' is at least as great as that for 'A is accepted, and not-A'. The latter formulation does not use the notion of truth; and so it suffices to explain what is meant by truth here.

9.2 NECESSARY CONDITIONS

If some people prefer that accepted hypotheses be false, they would violate the condition I stated in the preceding paragraph. I would say that such a person has unscientific values. Nevertheless, the person could satisfy all the axioms of Chapter 8 and hence have a cognitive utility function. In this case, the cognitive utility function does not represent scientific values.

Let a *scientific utility function* be any cognitive utility function that represents scientific values. That is, a scientific utility function is a cognitive utility function such that, if a person had it, we would say that person had scientific values. I do not assume that there is some one, essentially unique[2] scientific utility function. In fact, I am quite sure such a thing does not exist. But some kinds of values are clearly unscientific. In this section, I will propose four necessary conditions for a cognitive utility function to count as scientific. If these conditions are accepted, then any cognitive utility function that violates them is unscientific; but there can still be many, essentially different, cognitive utility functions that are scientific.

9.2.1 Respect for truth

The first necessary condition for a scientific utility function has already been mentioned: The cognitive utility of accepting a hypothesis when it is false cannot be higher than that of accepting the hypothesis when it is true. Letting $u(A, x)$ denote the cognitive utility of accepting A when the true state is x, we can formalize this condition as

$$u(A, x) \geq u(A, y), \quad \text{for all } A \in \mathbf{X},\ x \in A, \text{ and } y \in \bar{A}. \quad (9.1)$$

(For those who skipped Chapter 8: $\mathbf{X}$ is a set of subsets of the set of states X. Elements of $\mathbf{X}$ are thus sets of states, and are regarded as events or propositions.)

In Chapter 8, I assumed that the cognitive utility of accepting A when it is true is independent of which state in A obtains. That is,

$$u(A, x) = u(A, y), \quad \text{for all } A \in \mathbf{X} \text{ and } x, y \in A. \quad (9.2)$$

In this case, we can let $u(A, t)$ denote the cognitive utility of accepting A when it is true; that is, $u(A, t)$ will be the value of

---

[2]For those who read Chapter 8: What I mean by "essentially unique" is unique up to a positive affine transformation almost everywhere.

$u(A, x)$ for any $x \in A$. And then we can express (9.1) a little more perspicuously as

$$u(A, t) \geq u(A, x), \quad \text{for all } A \in \mathbf{X},\ x \in X. \qquad (9.3)$$

Because the inequality is weak, (9.3) does not say that a positive value is put on truth, only that a positive value is not put on falsehood. So (9.3) is consistent, for example, with the view that scientists care about whether their theories correctly describe observable phenomena but do not care whether they are correct about unobservable entities.[3] Although I suspect that scientists typically prefer to be right about unobservables as well as observables, I do not think that someone who is indifferent about correctly describing unobservables is thereby unscientific. For this reason, I would not require the inequality in (9.3) to be strict. More generally, there may be many things a person does not care whether they are right or wrong about, and this does not make a person unscientific; hence the weak inequality in (9.3) cannot be replaced by a strict one.

A number of philosophers of science have maintained that scientists do not value truth but do value agreement between theory and available evidence. Kuhn (1977, 1983) took this view, and similar positions have been taken by Laudan (1977), Ellis (1988), and others. Superficially, this position may appear to be in conflict with both (9.1) and (9.2). For might it not be the case that the evidence for a theory is better in some states in which the theory is false than in some states in which the theory is true? And might it not be the case that the evidence for a theory could vary in strength in different states in which the theory is true? If so, and if agreement with evidence is a scientific value, then we have a conflict with (9.1) and (9.2).

However, it must be remembered that acceptance is being understood here as acceptance of a total corpus (cf. Section 6.3.1). If the evidence is different in different states, and if that evidence is accepted, then what is accepted as corpus is different in the different states. But for a counterexample to (9.1) or (9.2), what is needed is a case in which the same proposition A is

---

[3]This view, which has been put forward by van Fraassen, will be discussed in Section 9.6.

accepted in two different states. Hence so long as we are dealing with accepted evidence, claims about cognitive utility being a function of the available evidence cannot be inconsistent with (9.1) or (9.2).

If the utility of accepting A were a function of the probability of A, then we could get counterexamples to (9.1) and (9.2). For example, suppose that accepting A when it has a high probability has higher utility,[4] other things equal, than accepting A when it has a low probability. Then the states will need to specify the probability of A, and if A is true in both states x and y but has a higher probability in the former, then $u(A, x) > u(A, y)$, in violation of (9.2).

While the situation envisaged would be a counterexample to (9.2), there is no reason to think accepting propositions with a high probability is part of the ends of science. To the extent that high probability is a reason for accepting a theory, this can be understood as being because high probability increases the expected utility of accepting a theory; we do not need to suppose in addition that it increases the actual utility of accepting a theory, when that theory is true. In other words, the desirability of high probability is adequately accounted for by taking it to be a means to the end of accepting true theories, and without taking it to be an end in itself.

Furthermore, if the utility of accepting a true theory covaried with the probability of that theory, we would get into the sort of difficulties discussed in Section 7.3. We would have that arbitrary shifts in probability could increase expected cognitive utility and that gathering cost-free evidence could reduce expected cognitive utility. Since both possibilities conflict with our views about what would be rational so far as scientific goals are concerned, we have a strong positive reason to deny that the utility of accepting a true theory varies with the probability of that theory. And more generally, we have positive reason to say that the utility of accepting a theory does not depend on the probability of that theory.

---

[4]'Utility' here is actual utility, not expected utility. The expected utility of accepting A will covary with the probability of A; but we are considering here the quite different claim that the actual utility of accepting A covaries with the probability of A.

So I will assume that both (9.1) and (9.2) are necessary conditions for a utility function to represent scientific values.

9.2.2 Respect for information

The concern for *informative* truths would clearly be violated if we did not have

$$u(AB, t) \geq u(A, t), \quad \text{for all } A, B \in \mathbf{X}. \tag{9.4}$$

Since the inequality is weak, a violation of this condition would be a situation in which a positive value is put on avoiding some information. For example, married persons may sometimes judge it preferable to be agnostic about their spouse's fidelity, rather than accept the existence of infidelity, even when there is indeed infidelity. This is not necessarily irrational (cf. Section 1.8); but it is unscientific.

Suppose a, b, and c are true statements. Then $abc \vee \bar{a}\bar{b}\bar{c}$ and $abc \vee \bar{a}bc \vee \bar{a}\bar{b}\bar{c}$ are both true statements, with the former being logically stronger than the latter; hence (9.4) entails that accepting $abc \vee \bar{a}\bar{b}\bar{c}$ should have cognitive utility at least as high as accepting $abc \vee \bar{a}bc \vee \bar{a}\bar{b}\bar{c}$. However, Tichý (1978) and Oddie (1981) have argued that in this example, it is better to accept $abc \vee \bar{a}bc \vee \bar{a}\bar{b}\bar{c}$ than to accept $abc \vee \bar{a}\bar{b}\bar{c}$. Their reasoning is that in accepting $abc \vee \bar{a}\bar{b}\bar{c}$ one stands a 50 percent chance of being as far from the truth as possible, while accepting $abc \vee \bar{a}bc \vee \bar{a}\bar{b}\bar{c}$ gives one a 2/3 chance of being close to the truth and only a 1/3 chance of being as far from the truth as possible.[5] If Tichý and Oddie were right about this, then we would have a counterexample to (9.4). But they are not right. Acceptance of a disjunction is acceptance of the disjunction, not acceptance of a randomly selected disjunct. So, in particular, acceptance of $abc \vee \bar{a}\bar{b}\bar{c}$ does not give one any chance of accepting $\bar{a}\bar{b}\bar{c}$, and hence it does not give one any chance of being as far from the truth as possible. Thus we have not yet been given a good reason to reject (9.4) as a necessary condition for a scientific utility function. (For a more extended discussion of the position of Tichý and Oddie, see [Niiniluoto 1987, sec. 6.6].)

---

[5] These judgments of closeness to the truth depend on Tichý's measure of verisimilitude, which will be presented in Section 9.4.2.

213

The weak inequality in (9.4) should not be made strict. For one thing, A might entail B, in which case AB expresses the same proposition as A, and so identity must hold in (9.4). But even if A does not entail B, it need not be unscientific to have $u(AB, t) = u(A, t)$. For example, let A specify the age of the universe and B specify this week's top-selling record; I don't care whether I know B, and so for me $u(AB, t) = u(A, t)$, even though A does not entail B; but this is surely not a reason to say my values are unscientific. For another sort of example, suppose (as in the example of Section 6.3.2) that we are interested in the value of some real-valued parameter, and that

$$u(A, t) = \frac{k}{1 + \sup(A) - \inf(A)}$$

for some k. Then setting $A = [0, 1]$ and $B = \{r : r \neq 1\}$, we have

$$u(AB, t) = u(A, t) = k/2,$$

even though A does not entail B. But I would not want to call someone unscientific, merely because they have this utility function.

On the other hand, someone who does not care about being right about anything would not have scientific values. Thus for a person with scientific values, there must be some proposition A such that the person thinks that accepting A when it is true is better than complete agnosticism on empirical matters. Formally, this amounts to saying that the following condition is necessary for scientific values:

$$u(A, t) > u(X, t), \quad \text{for some } A \in \mathbf{X}. \tag{9.5}$$

9.2.3 Impartiality

If science is a disinterested search for truth, then the scientific utility of accepting the whole truth would be the same, whatever that truth may be. Formally, this condition is that

$$u(\{x\}, t) = u(\{y\}, t), \quad \text{for all } x, y \in X. \tag{9.6}$$

This condition requires that the utility of accepting the whole truth be the same in every state, and hence that the graph of the utility of this act be a horizontal straight line.

But is science a disinterested search for truth, in this sense? It is often said that simplicity is a goal of science. This claim can be interpreted in different ways, and on one interpretation it is inconsistent with (9.6). I will therefore mention here the main interpretations and discuss their compatibility with (9.6).

Some writers, such as Jeffreys (1931), have held that scientists prefer simple theories because these have a higher probability of being true. On this view, simplicity is valued not because it is one of the ends of science, but rather because it is viewed as a means to the end of having true theories. Hence this view is consistent with the idea that the cognitive utility of accepting a complete true theory does not depend on the simplicity of the theory; in other words, the view is consistent with (9.6).

Popper (1959), Kemeny (1955), Sober (1975), and Rosenkrantz (1977) have all held that the simplicity of a theory covaries with the content of the theory. On this view too, simplicity is a means to the goal of having true informative theories, not a separate end of science. So this view of the role of simplicity in science is also consistent with (9.6).

Finally, there are those, such as Kuhn (1977, 1983), who see simplicity as an ultimate goal of science, not a means to the goal of having true informative theories. For these, the utility of accepting a complete true theory will vary, depending on the simplicity of the theory (and, in Kuhn's case, other factors also). So if this view is accepted, (9.6) would have to be rejected as a condition on a scientific utility function.

This is not the place to attempt a definitive account of the role of simplicity in science. However, I will express my sympathies with the view that simplicity is not an ultimate end of science, but only a means to the goal of having true informative theories. I also note that scientists themselves often talk in a way that fits better with the idea that simplicity is valued as an indicator of truth, rather than an end in itself. For example, biologist Francis Crick writes, "Elegance and a deep simplicity, often expressed in a very abstract mathematical form, are useful guides in physics, but in biology such intellectual tools can be very misleading" (Crick 1988, p. 6). A similar statement, by the physicist John Schwarz, is quoted in (Kaku and Trainer 1987, p. 195). These scientists evidently value simplicity only so

215

long as it appears to be a reliable guide; hence it is for them a means to an end, not an end in itself. Going back further, Newton's invocation of simplicity is supported with the claim that it is an indicator of truth: "Nature is pleased with simplicity, and affects not the pomp of superfluous causes" (Newton 1726, vol. II, p. 398).

Consequently, while I acknowledge that simplicity plays a role in theory choice, I hold that it is not an intrinsic goal of science; and hence I hold that (9.6) really is a necessary condition for a utility function to be scientific. Furthermore, reducing the number of postulated ends of science is surely a gain in simplicity, so the advocates of simplicity would themselves seem committed to agreeing that simplicity should not be taken to be an end of science.

9.2.4 Contradiction suboptimal

Conditions (9.3)–(9.6) impose no restrictions on preferences regarding acceptance of the contradictory proposition $\emptyset$. (In the case of (9.3)–(9.5), this is because there is no such cognitive consequence as $< \emptyset, t >$.) But someone who weakly prefers to accept a contradiction than to accept the complete truth is surely being unscientific, if not irrational. Hence a further condition satisfied by scientific utility functions is

$$u\left(\{x\}, t\right) > u(\emptyset, y), \quad \text{for all } x, y \in X. \tag{9.7}$$

I would not claim that conditions (9.3)–(9.7) jointly constitute sufficient conditions for a cognitive utility function to be scientific; though I am not currently aware of any further conditions that seem plausible to me. But I doubt that any reasonable conditions on a scientific utility function could be so strong as to identify an essentially unique utility function as scientific.

9.3 VALUE INCOMMENSURABILITY

Kuhn (1977) has maintained that certain values are partly constitutive of science, but that these values can be interpreted differently by different scientists. He also says the weight to be given to each can be assessed in different ways without being unscientific. In (1962), Kuhn seemed to be arguing from these

216

premises to the conclusion that scientific debates cannot be rationally persuasive. For example, he wrote:

To the extent, as significant as it is incomplete, that two scientific schools disagree about what is a problem and what a solution, they will inevitably talk through each other when debating the relative merits of their respective paradigms. In the partially circular arguments that regularly result, each paradigm will be shown to satisfy more or less the criteria that it dictates for itself and to fall short of a few of those dictated by its opponent. (pp. 109f.)

Doppelt (1978) has elaborated and defended this argument that divergent values entail the incommensurability of theories.

While I differ from Kuhn on what the ultimate scientific values are, I agree with him that values necessarily play a role in theory choice, and that different scientists can have different values without being unscientific. Am I then committed to the view that scientific debates cannot be rationally persuasive? I am not.

It is true that if two scientists have different values, then they can disagree about the relative merits of different theories, without either having made a mistake. For example, suppose the utility of accepting theory A when it is true is 0.6 for scientist 1, and 0.4 for scientist 2; and suppose further that the utility of accepting A when it is false is -0.5 for both. Then if both agree that the probability of A is 0.5, accepting A will have a positive expected utility for scientist 1 and a negative expected utility for scientist 2. Assuming that the expected utility of suspending judgment is zero, this means that scientist 1 prefers accepting A to suspending judgment, while scientist 2 has the opposite preference. And neither can be said to be wrong; their difference simply reflects different values.

But it does not follow from this that the difference between these scientists cannot be resolved by rational means. As further evidence is acquired, the probability of A will shift. If it rises to 0.6, then scientist 2 will come to agree with scientist 1 that accepting A is preferable to suspending judgment. Similarly, if the probability of A falls to 0.4, then scientist 1 will adopt scientist 2's preferences. Thus it is possible for the difference between these scientists to be resolved by gathering further evidence. This is a rational way to resolve their difference and

217

does not require either party to adopt the values of the other.[6]

Of course, there is no guarantee that the requisite evidence will be found to settle scientific differences resulting from different values. But the history of science indicates that differences usually do get settled, and this happens primarily as a result of additional evidence being acquired, not by resolving divergent values. On the other hand, at the cutting edge of research there always are disagreements, and we should not require a philosophy of science to deem all such disagreements as irrational or unscientific. In fact, as Kuhn (1977) has persuasively argued, scientific progress is enhanced by the fact that different scientists sometimes accept different theories. On the account I am offering, such differences may be due in part to different probabilities on the part of the scientists involved, but may also be due to a difference in their values, even though those values are scientific.

9.4 VERISIMILITUDE

In Chapters 6 and 8, I allowed for the possibility that scientists would rather a false accepted theory be close to the truth than not so close. But what does it mean for a false theory to be more or less close to the truth? This notion cannot be explicated in the way I explicated the notion of truth itself. Furthermore, skepticism about the meaningfulness, or usefulness, of this notion has often been expressed. I will now argue that the notion is worth trying to make sense of, and that the representation theorem of Chapter 8 provides a way of making sense of it.

9.4.1 Why verisimilitude?

The idea that false theories can be more or less close to the truth has often been invoked by those who see science as aiming at truth. An important motivation has been to give an account of *progress* in science. The history of science is a history of false

---

[6]Subsequently, Kuhn himself has noted that evidence can produce agreement amongst scientists with different algorithms or values (Kuhn 1977, p. 329). He does still maintain that change in theory produces a change in values, but notes that this "does not make the decision process circular in any damaging sense" (p. 336). Thus before Doppelt's article was published, Kuhn had repudiated the thesis Doppelt attributed to him.

theories, and yet we want to say that science is making progress. Consequently, those who take science to be aiming at truth are apt to suggest that science is making progress in that it is tending to approach truth more and more closely, even though all past and present scientific theories may be strictly false.[7] According to this view, Newtonian mechanics is an advance over Aristotelian mechanics because, despite the fact that both these theories are false, Newtonian mechanics is closer to the truth.

But does a realist really need to take this approach? An alternative suggestion (made by Giere [1983, sec. 9]) is that scientists should not accept that their theories are literally correct, but rather that reality is like their theoretical models in certain specified respects, to within certain specified degrees of error. If the qualifications here are chosen suitably, what is accepted will stand a good chance of being true. One could then say that progress in science consists in this sequence of generally true propositions becoming more informative over time.

There are several problems facing this latter approach. The first is that scientists do not appear to have put limits on what they accepted that were sufficiently broad to make it the case that what they accepted was true. For example, the historical evidence strongly suggests that Newtonians accepted that gravitation acts instantaneously over the whole universe, that the mass of bodies is completely independent of their velocity, and so on. No margins of error for these theories were specified; and if the scientists had some tacit margins in mind, there is no reason to think these were broad enough that what the scientists accepted was true by our lights (i.e., consistent with relativity theory). So Giere's approach is not able to prevent the conclusion that Newtonians accepted false propositions. Since we nevertheless want to say that Newtonianism represented progress over Aristotelianism, we still have a motivation for saying that some false theories are closer to the truth than others.

Defense of the progressiveness of science is not the only reason for allowing that false theories can be more or less close to the

---

[7]This view of the nature of scientific progress has been expressed by Popper (1963, p. 231), Niiniluoto (1984, ch. 5), and Newton-Smith (1981), among others.

219

truth. In many cases, the notion is intuitively compelling in its own right. Consider, for example, Cavendish's hypothesis:

H_C: The electrostatic force falls off as the nth power of the distance, for some n between 1.98 and 2.02.

This hypothesis is false if the true value of n is anything less than 1.98 or greater than 2.02; but we would certainly be inclined to say that Cavendish was closer to the truth if the true value of n was 2.021 than if it was 3.

9.4.2 A little history

There is, then, good motivation for saying that false theories can be more or less close to the truth. But we can say this meaningfully only if the required notion of distance from truth is meaningful. On this point, there is considerable skepticism; for example, Laudan (1977, pp. 125f.) writes that "no one has been able even to say what it would mean to be 'closer to the truth.'" Laudan's complaint is not that nobody has tried to say what it means; and in fact, the literature now contains a large number of attempts to say what it means. But there is a widespread sense that these attempts have not been successful, and this has fueled skepticism about the possibility of making clear sense of the notion.

The first rigorous account of 'closeness to the truth' was given by Popper (1963, pp. 231–7, 391–403; 1972, pp. 47–60). The notion of closeness to truth that interested Popper was closeness to the *whole* truth; thus a true but uninformative statement is not close to the truth, in this sense. Popper introduced the term *verisimilitude* to refer to closeness to the truth in this sense. Thus the verisimilitude of a theory is the degree to which it approximates the ideal of entailing all truths and no falsehoods. Popper observed that if all true consequences of B are also entailed by A, and if any false consequence of A is also entailed by B, the verisimilitude of A is at least as great as that of B. If in addition A entails some true statements not entailed by B, or B entails some false statements not entailed by A, then Popper said the verisimilitude of A is greater than that of B. These conditions on verisimilitude have come to be called Popper's "qualitative definition of verisimilitude."

Unfortunately, this qualitative definition permits the verisimilitude of theories to be compared only in special cases. To compare the verisimilitude of A and B, it must be the case that either (i) all true consequences of B are also entailed by A, and any false consequence of A is also entailed by B; or (ii) this relation must hold with A and B interchanged. However, Tichý (1974) and Miller (1974) showed that no two false theories satisfy these conditions, and hence no two false theories are comparable using Popper's qualitative definition of verisimilitude. But a central motivation for introducing the notion of verisimilitude is to say that some false scientific theories are closer to the truth than earlier ones.

Popper also gave a quantitative definition of verisimilitude, which was intended to define the verisimilitude of all statements. Letting H_t be the conjunction of the true consequences of H, and p a logical probability function, Popper suggested that the *truth content* of H could be taken to be measured by $1 - p(H_t)$. He also suggested that the *falsity content* of H could be measured by $1 - p(H|H_t)$. Thinking of the verisimilitude of a statement as its truth content minus its falsity content then leads to the definition[8]

$$v(H) = p(H|H_t) - p(H_t). \tag{9.8}$$

One problem with (9.8) is that it rests on the assumption that there is such a thing as logical probability. Ramsey had already argued cogently against the existence of such a thing in (1926), and Carnap's later heroic efforts to make sense of the notion in the end (1952) only confirmed that Ramsey had been right.[9] Nor could we substitute subjective for logical probability in (9.8), since that would make it impossible to be certain of the truth of any hypothesis with positive verisimilitude. As Popper himself insisted, verisimilitude is not meant to be an epistemic notion.

Even if the required logical probability function existed, (9.8) would still be an unsatisfactory definition for other reasons.

---

[8]For numerical calculations of verisimilitude, Popper prefers to divide the following definiendum by the normalization factor $p(H|H_t) + p(H_t)$ (Popper 1963, p. 399; 1972, p. 334). But this is a refinement that need not concern us here.

[9]In a penetrating recent study, van Fraassen (1989, ch. 12) reaffirms this conclusion.

Note that a statement is a true consequence of H just in case it is a consequence of $H \vee T$, where T is the true complete theory (relative to some partition). Hence we can substitute $H \vee T$ for H_t in (9.8), obtaining

$$
\begin{aligned}
v(H) &= p(H|H \vee T) - p(H \vee T) \\
&= \frac{p(H \vee HT)}{p(H \vee T)} - p(H \vee T).
\end{aligned}
$$

If H is false, then H and T are disjoint, and we get

$$
v(H) = \frac{p(H)}{p(H) + p(T)} - p(H) - p(T).
$$

Thus the verisimilitude of false theories is a function of the probability of the theory. In particular, all false theories with the same probability have the same verisimilitude. As an example of this, suppose an urn contains 100 balls, each of which is either black or white. Proponents of logical probability would say that (in the absence of other information) each attribution of colors to balls has the same logical probability. Then if all the balls are in fact white, Popper's account deems the hypothesis that all the balls are black just as close to the truth as the hypothesis that balls 1–99 are white and ball 100 is black. But this is not the judgment we would make. (Tichý [1974] made essentially this point.)

Although Popper's attempts to define verisimilitude were unsuccessful, their failure prompted a host of other attempts to achieve Popper's goal: to define a mathematical function that would be a measure of the verisimilitude of a hypothesis. Tichý made a start at the end of his 1974 paper criticizing Popper's approach; since this proposal was relatively simple, and since many later proposals are generalizations or modifications of the sort of approach Tichý adopted, it will be worthwhile to briefly review this first proposal of Tichý's.

Tichý considered a language in which there are finitely many primitive sentences, and all other sentences in the language are truth functional combinations of these sentences. (So the language belonged to what is called "propositional logic.") For a simple example, let us suppose there are just three primitive sentences, a, b, and c. The maximally specific sentences that can be expressed in this language are the eight sentences of the form abc, $ab\bar{c}$, ..., $\bar{a}\bar{b}\bar{c}$. These are called *constituents*, and they are the linguistic equivalent of states of nature. In particular, the constituents form a partition: Exactly one of them must be true. Tichý proposed that the *distance* of a constituent from the truth could be taken to be measured by the number of primitive sentences that the constituent is wrong about. For example, if a, b, and c are all true, the distance of abc from the truth would be 0; the distance of $\bar{a}bc$ from the truth would be 1; the distance of $\bar{a}\bar{b}c$ from the truth would be 2; and so on.

That definition only defines distance from truth for constituents. It remains to extend the definition to other sentences. Tichý noted that for languages of the kind here considered, any sentence that can be expressed in the language is equivalent to a disjunction of constituents. For example, the sentence ab is equivalent to $abc \lor ab\bar{c}$. The disjunction of constituents that is equivalent to a given sentence is called the *disjunctive normal form* of that sentence. Tichý proposed that the distance from the truth of any sentence in the language could be taken to be the average distance from the truth of the constituents in the disjunctive normal form of the sentence.[10] For example, the distance of ab from the truth, on this proposal, would be $(0 + 1)/2 = 1/2$.

The notion of distance from truth, that Tichý intended to explicate, is distance from the *whole* truth (where the whole truth is represented by the true constituent). Distance from the truth is thus the inverse of the notion of verisimilitude. So given a definition of distance from truth, we can define verisimilitude as a function that varies inversely with distance from truth. Thus Tichý's definition of distance from truth also provides a definition of verisimilitude.

How satisfactory is Tichý's definition, within its intended domain of application? Miller (1974, sec. 6) pointed out that on Tichý's definition, the verisimilitude of one and the same statement may be different, depending on what the primitive sentences of the language are taken to be. Tichý denies that this

---

[10]Tichý (1974, p. 159) writes that this quantity is to be taken as the *verisimilitude* of the sentence; but this is a slip.

is the case. His argument (Tichý 1978, sec. 8) seems to assume that two sentences do not express the same statement unless both occur in a single language, and definitions that entail the equivalence are adopted. This is surely false. 'Der Schnee ist weiss' in German expresses the same statement as 'Snow is white' in English, though neither sentence occurs in the other language. (Quotation names of each sentence occur in the other language, but the sentences themselves do not.)[11]

But even if Tichý were right about this, there would remain a problem of arbitrariness in Tichý's account. On Tichý's account, the verisimilitude of a sentence uttered by a scientist depends on what we take the scientist's language to include, and on which sentences in that language are taken to be primitive. If these choices of language and primitive sentences are arbitrary (and Tichý has given no reason to think they are not), then the verisimilitude to be attributed to a sentence uttered by a scientist will also be a matter of arbitrary choice. Such arbitrariness is unacceptable because an important motivation for introducing the notion of verisimilitude in the first place was to be able to say that scientific theories are getting closer to the truth. If the verisimilitude of those theories is arbitrary, then the claim that science is getting closer to the truth is equally arbitrary.

9.4.3 The case for a subjective approach

Although we have no use for an arbitrary conception of verisimilitude, it is a mistake to think that verisimilitude can be independent of human interests or values. Every false theory is like the truth in some respects and unlike it in others; so to say that it is like (or unlike) the truth overall requires a judgment as to which respects are important. If this judgment is not to be arbitrary, it must be founded on the degree to which we value correctness of various kinds.

Many philosophers of science seem intent on banishing value judgments from science; and so if persuaded that verisimilitude rests on value judgments, they would banish it too. (This seems to be the position of Urbach [1983].) But even if verisimilitude

---

[11] For another criticism of Tichý's position, see (Urbach 1983, sec. 1).

224

were banished, value judgments would remain. If one concedes that science aims to have theories that are both true and informative, then comparison of theories requires weighing the relative importance of these different desiderata. For example, if A is logically stronger than B, then acceptance of A would give more information than B, but also runs a greater risk of error. The decision of which to accept (as corpus) thus requires weighing the two desiderata against each other. If this weighting is not to be arbitrary, it must reflect our values.

We might also ask: What does it mean to "banish" verisimilitude? Presumably it means that the value of accepting a theory when it is false is independent of which aspects of the theory are false. But this is no less a value judgment than the contrary claim, that the value of accepting a false theory may be different depending on what aspects of the theory are false. So "banishing" verisimilitude is really just a different proposal about scientific values, and does not achieve the desired aim of banishing value judgments from science.[12]

Given that judgments of verisimilitude must depend on our values, we might fix the problem of language-relativity in Tichý's account of verisimilitude by saying that the choice of a language is fixed by our values. This has been proposed by Niiniluoto (1987, p. 459). The problem of arbitrariness is thereby solved, since it is no longer the case that any language may be used for assessing verisimilitude using Tichý's measure.

However, once we have allowed that judgments of verisimilitude are relative to our interests, I see no reason to think that those interests can be always be represented by a fixed measure on a suitably chosen language. In any event, the use of syntax to define verisimilitude seems unnecessarily indirect. A more direct route would be to have the measure defined on propositions (which can be the same, though expressed by different sentences) and let our interests be reflected in the choice of the measure.

---

[12]The aim might seem to be more nearly approximated by Koertge's (1979) suggestion that we need not make judgments about the relative merits of false theories. This suggestion gives up on trying to say that science has been making progress even though its theories have been false. But even if one were willing to abandon this view of scientific progress, Koertge's suggestion would still be untenable. For the rationality of accepting a theory depends, in part, on the value of accepting the theory when it is false.

225

Even though verisimilitude is relative to our interests, it could still be objective, in the sense that everyone has the same interests and hence verisimilitude is the same for everyone. But on the face of it, this is implausible. Individuals differ in the relative values they place on everything else; why not also on the relative importance of content and truth and on the relative importance of the different possible ways of being wrong?

There is an analogy here with early work in utility theory. At first it was assumed that the value of a bet was measured by its expected monetary return. This assumption was refuted by Nikolaus Bernoulli's "St. Petersburg paradox," where a game with infinite expected monetary value is worth very little. Daniel Bernoulli (1738) proposed that the value of a bet is measured by its expected *utility*, and argued that utility is a logarithmic function of money. At this time, it was still assumed that there is some one function that expresses the utility of money.[13] But nobody thinks that any more; we now allow that different people may have different utility functions for money.

A measure of verisimilitude is essentially a measure of the cognitive utility of accepting a hypothesis under different circumstances. The attempts of Popper, Tichý, and others to produce a mathematical definition of verisimilitude are thus analogous to Daniel Bernoulli's attempt to define the utility of money. We gave up on Bernoulli's project long ago, and I think it is time we gave up on the Popperian analog of that project.

I began this chapter by distinguishing scientific values from other values, and in Section 9.2 I suggested some conditions that the former satisfy. But this still allowed that different sets of values can count as scientific, just as different sets of values can count as unscientific. Judgments of verisimilitude require a weighting of the relative importance of different ways of being wrong; and I see no reason to think that science is so narrowly defined that it can uniquely determine these weightings.

Recall that Popper's original motivation for proposing his definitions of verisimilitude was to show that the notion of

---

[13] Daniel Bernoulli did allow that there were some "exceedingly rare" exceptions to his rule. His example is a prisoner who needs a fixed sum of money to purchase freedom. But even here, the utility function is conceived as fixed by the person's external circumstances.

226

verisimilitude was meaningful. However, we do not need there to be one unique verisimilitude function in order for the notion of verisimilitude to be meaningful. Consider again the analogous problem of interpreting the notion of the utility of money. On the approach advocated in Chapter 1, the utility of money is relative to persons; and a person has a particular utility function for money if the person's preferences regarding monetary gambles maximize expected utility relative to that utility function (and some probability function). Applying the same approach to verisimilitude, we will say that verisimilitude is relative to persons; and for a person to have a particular measure of verisimilitude is for the person to have a particular kind of cognitive utility function. Further, a person has a particular kind of cognitive utility function if the person's preferences regarding cognitive gambles maximize expected cognitive utility relative to that cognitive utility function (and some probability function). In this way, we can give meaning to the notion of verisimilitude, without requiring that there be some unique measure of verisimilitude that is the same for everyone. In the next subsection, I will fill in the details of how this subjective approach to verisimilitude may be carried out.

9.4.4 A subjective definition of verisimilitude

The representation theorems of Chapter 8 show that rational persons have cognitive preferences that are representable by a probability function and a cognitive utility function (though these functions need not be unique). On the favored interpretation of Theorem 8.2, the cognitive preferences are over a set D of cognitive options that consists of (i) countable experiments, and (ii) the act f_T of accepting the complete truth (relative to the chosen partition of states). Recall that the act of accepting a hypothesis is a degenerate case of a countable experiment (Section 8.3.2), and hence the representation applies in particular to acts of accepting a hypothesis. Furthermore, Theorem 8.2 allows the cognitive consequence of accepting A when A is false to be different for every $x \notin A$, and hence it allows the cognitive utility of accepting A to be different for every state in $\bar{A}$. What I will now do is show how such a cognitive utility function can

227

be used to define a subjective notion of verisimilitude. I will assume that we are dealing with a person whose cognitive utilities satisfy (9.3)–(9.7).

A simple subjective measure of verisimilitude could be obtained by identifying the verisimilitude of a hypothesis with the cognitive utility of accepting it. But such a measure would be unsatisfactory. Cognitive utilities are, at best, unique only up to a positive affine transformation (as we saw in Chapter 8). This simple definition of verisimilitude would thus give different values of verisimilitude for one and the same hypothesis, depending on which of the equivalent utility functions we chose to use.

We can avoid this difficulty by normalizing the cognitive utility function. It is natural to do this in such a way that the verisimilitude of the complete true theory[14] is 1, while that of a tautology is 0. This leads to the following definition. Here $v_x(A)$ denotes the verisimilitude of proposition A when x is the true state. By (9.6), $u(\{x\}, x)$ is the same for all $x \in X$, and I will denote this value by u_T. Also, by (9.2), $u(X, x)$ is the same for all $x \in X$; this value will be denoted u_X. The definition of verisimilitude is then

Definition 9.1.

$$v_x(A) = \frac{u(A, x) - u_X}{u_T - u_X}.$$

It follows immediately from this definition that for all $x \in X$, $v_x(\{x\}) = 1$ and $v_x(X) = 0$.

Anything deserving the name verisimilitude must satisfy the condition that a hypothesis is at least as close to the truth when it is true as when it is false. Formally, what we should have is

$$v_x(A) \geq v_y(A), \quad \text{for all } x \in A, \ y \in \bar{A}. \tag{9.9}$$

[14]The "complete true theory" is the theory that asserts that the true state obtains. That is to say, it asserts that the true element of some partition, chosen by us, obtains. Each element of this partition is supposed to be sufficiently specific that it, together with the chosen act, determines everything that is of value (Section 1.1); hence the complete true theory, in a cognitive decision problem, should specify the truth value of every proposition whose truth value we would like to know. However, it need not be complete in any stronger sense than this.

From (9.4) and (9.5) we have that $u_T - u_X > 0$; and this together with (9.3) and Definition 9.1 entails (9.9).

Verisimilitude is meant to represent closeness to the *whole*
truth. So if A and B are both true, the verisimilitude of AB
should be at least as great as that of A. Formally:

$$v_x(AB) \geq v_x(A), \quad \text{for all } x \in AB. \quad (9.10)$$

This inequality follows immediately from (9.4) and Definition 9.1.

Let x^* denote the true state. Then the actual verisimilitude
of any proposition A is $v_{x^*}(A)$, which I will write simply as
$v(A)$. At this point, I have addressed Laudan's complaint, mentioned earlier, that "no one has been able even to say what it
would mean to be 'closer to the truth'." The verisimilitude of a
hypothesis has been defined partly in terms of a person's cognitive utilities (and hence preferences), and partly in terms of
the true state of nature.

The fact that the definition of verisimilitude refers to the true
state x^* brings me to a second complaint that Laudan has made
about verisimilitude, namely that no one has been able "to offer criteria for determining how we could assess" verisimilitude
(Laudan 1977, p. 126; see also 1981, p. 32). Now it is true that
we do not in general know which state is x^*, and it follows from
this that we generally do not know the value of the verisimilitude of a theory. But we can calculate an *expected* value for the
verisimilitude of a theory. Assuming X is countable, we have,
for any proposition A,

$$\mathcal{E}[v(A)] = \sum_{x \in X} p(x)v_x(A).$$

As a simple example of this, let $\{H_1, H_2, H_3\}$ be a partition
of hypotheses. Suppose that

$$u(H_1, x) = \begin{cases} 0.5 & \text{if } x \in H_1 \\ 0.25 & \text{if } x \in H_2 \\ -0.5 & \text{if } x \in H_3. \end{cases}$$

Suppose also that the utility of correctly accepting that state
x is the true state is 1, for all x, and the utility of accepting X

229

(i.e., completely suspending judgment) is 0. Then Definition 9.1 gives that $v_x(H_i) = u(H_i, x)$, for all x and i. The expected verisimilitude of H_1 is thus

$$\mathcal{E}[v(H_1)] = 0.5p(H_1) + 0.25p(H_2) - 0.5p(H_3).$$

For instance, if all the H_i have the same probability, then the expected verisimilitude of H_1 is $1/12$.

Propositions with maximum expected verisimilitude are also those whose acceptance has maximum expected cognitive utility.

Proof.

$$\mathcal{E}[v(A)] \quad = \quad \mathcal{E}\left[\frac{u(A,x^*)-u_X}{u_T-u_X}\right], \qquad \text{by Definition 9.1}$$

$$= \quad \frac{\mathcal{E}[u(A,x^*)]-u_X}{u_T-u_X}, \qquad \text{since } u_T \text{ and } u_X \\ \text{are constants.}$$

Since $u_T - u_X > 0$, it follows that $\mathcal{E}[v(A)]$ is maximized by all and only those propositions A that maximize $\mathcal{E}[u(A, x^*)]$. Since $\mathcal{E}[u(A, x^*)]$ is the expected utility of accepting A, this is the desired result.

From this result and the principle of maximizing expected utility, we have that rationality requires a person with a scientific cognitive utility function to accept a proposition that maximizes expected verisimilitude.

If all the axioms of Chapter 8 are satisfied, then Theorem 8.2 guarantees that the utility function is unique, up to a positive affine transformation, almost everywhere. Consequently, verisimilitude, as defined here, will be unique almost everywhere. (Definition 9.1 ensures that utility functions differing only by a positive affine transformation yield the same measure of verisimilitude.) And the fact that this uniqueness only holds almost everywhere is not significant, since expected verisimilitude is unaffected by differences on a set of probability 0. However, I have taken the position that the connectedness axiom (Axiom 1) is not a requirement of rationality. Consequently, I allow that the utility functions we can attribute to a scientist need not

230

satisfy the above uniqueness condition. It will then follow from my definition of verisimilitude that the scientist's measure of verisimilitude is not uniquely determined, not even almost everywhere. Rather, the scientist's measure of verisimilitude will be represented by a set of verisimilitude functions, one for each utility function in some p-u pair in the scientist's representor (cf. Section 1.5). This is not, I think, a disturbing situation; there is a sense of unreality about measures of verisimilitude that purport to assign precise numbers as the verisimilitude of every hypothesis in every state.

9.5 INFORMATION AND DISTANCE FROM TRUTH

In Section 6.3, I suggested that scientists aim to accept hypotheses that are both informative and true – goals that tend to compete with one another. These two goals are both contained in the notion of verisimilitude, since verisimilitude is closeness to the *whole* truth. But the explication of verisimilitude given in the preceding section did not isolate these components of verisimilitude. The present section will make good that omission and show how measures of information and distance from the truth may be defined. I will also compare my definition of information with others in the literature.

9.5.1 A subjective definition of information

Information, as we are concerned with it here, is the extent to which a proposition answers a person's cognitive questions. Thus a measure of information is a measure of the relative importance of different questions to the person. As such, a measure of information must be a function of the person's values; and since different people can have different values, we should not require there to be one unique measure of information that is the same for everyone. Information, like verisimilitude, is understood here as a subjective notion. And in fact, the approach I will follow is to define informativeness in terms of verisimilitude. I assume throughout this section that the underlying cognitive utility function is scientific; that is, it satisfies (9.3)–(9.7).

If A is true, then the verisimilitude of A can itself be taken as a measure of the informativeness of A. So letting $c(A)$ denote

231

the information content of A, we can for true hypotheses set

$$c(A) = v(A). \tag{9.11}$$

For example, this identity entails that $c(\{x^*\}) = v(\{x^*\}) = 1$, and $c(X) = v(X) = 0$. This is as it should be, since $\{x^*\}$ is fully informative, while X conveys no information at all.

As for false propositions, their information content can be identified with what their verisimilitude would be if they were true. The verisimilitude A would have if it were true is $v_x(A)$, where x is any state in A. Of course, this makes sense only if A is not the empty (or contradictory) proposition $\emptyset$. But for $\emptyset$ we can simply adopt the convention that its informational content is 1. So my general definition of information is

Definition 9.2.

$$c(A) = \begin{cases} v_x(A) & \text{if } A \neq \emptyset \text{ and } x \in A \\ 1 & \text{if } A = \emptyset. \end{cases}$$

Any satisfactory conception of information must satisfy the condition

$$c(AB) \geq c(A), \quad \text{for all } A, B \in \mathbf{X}.$$

This condition follows from Definition 9.2.

Proof. If $A \neq \emptyset$ and $B \neq \emptyset$,

$$\begin{aligned} c(AB) &= v_x(AB), \quad \text{for } x \in AB \\ &\geq v_x(A), \quad \text{for } x \in AB, \text{ by (9.10)} \\ &= c(A), \quad \text{by Definition 9.2.} \end{aligned}$$

If $A = \emptyset$ or $B = \emptyset$, $c(AB) = 1$, and so again $c(AB) \geq c(A)$.

In Section 6.3.2, I discussed an example of estimating the true value of some real-valued parameter. Here the propositions A were identified with sets of real numbers; and I defined a measure of the content of any proposition A by setting

$$c(A) = \begin{cases} \frac{1}{1+\sup(A)-\inf(A)} & \text{if } A \neq \emptyset \\ 1 & \text{if } A = \emptyset. \end{cases}$$

I also defined the distance of A from the true parameter value r to be

$$d_r(A) = \begin{cases} \frac{\inf_{x \in A} |x-r|}{1+\inf_{x \in A} |x-r|} & \text{if } A \neq \emptyset \\ 1 & \text{if } A = \emptyset. \end{cases}$$

And I considered cognitive utility functions of the form

$$u_k(A, r) = kc(A) - d_r(A). \tag{9.12}$$

The function c in this example satisfies Definition 9.2.

Proof. Let $r \in A$. Then

$$
\begin{aligned}
c(A) &= \frac{1}{k}\left[u_k(A, r) + d_r(A)\right], \quad \text{by (9.12)} \\
&= \frac{1}{k}u_k(A, r), \quad \text{since } d_r(A) = 0 \text{ for } r \in A \\
&= \frac{u_k(A, r) - u_k(X, t)}{u_k(\{x\}, t) - u_k(X, t)}, \quad \text{since } u_k(X, t) = 0 \\
&\qquad \text{and } u_k(\{x\}, t) = k, \text{ for any } x \\
&= v_r(A), \quad \text{by Definition 9.1.}
\end{aligned}
$$

This shows that $c(A)$ satisfies Definition 9.2 when $A \neq \emptyset$. And $c(\emptyset)$ is defined to be 1, as Definition 9.2 requires.

To summarize this subsection: The concept of information that is relevant to the acceptance of propositions is one that measures the degree to which a proposition answers a person's cognitive questions. When the person's cognitive values are scientific, we can interpret this sense of information as the verisimilitude a proposition would have if true. Since verisimilitude is defined in terms of the person's cognitive utility function, which is in turn defined in terms of the person's cognitive preferences, information has here been defined ultimately in terms of the person's cognitive preferences. Of course, if the person's preferences are not complete, then this subjective measure of information, like that of verisimilitude, will be to some degree indeterminate – which is as it should be.

9.5.2 Comparison with other accounts of information

Measures of the informativeness of propositions have been discussed by a number of philosophers of science, since many have agreed that informativeness is part of the goal of science. As Bar-Hillel and Carnap (1953) observed, the relevant notion of information for this purpose is a semantic one, and thus the purely syntactic measures of Shannon (1949) do not measure what we are interested in here.

The semantic measures that have been proposed have, almost without exception, been based on the intuition that the informativeness of propositions varies inversely with their probability. For example, Bar-Hillel and Carnap (1953) proposed that the amount of information conveyed by a proposition A could be defined either by

$$c(A) = 1 - p(A) \tag{9.13}$$

or else by

$$c(A) = -\log_2 p(A) \tag{9.14}$$

where p is some suitable probability function. Virtually all other authors who have discussed measures of informativeness have accepted some variant of (9.13) or (9.14).[15] For example, Popper (1963, p. 392) uses (9.13).

As Hempel and Oppenheim observed in the conclusion of their classic (1948) paper, a crucial problem for definitions such as these is the problem of selecting, from among an infinity of formal possibilities, some suitable probability measure p. The usual suggestion was that p should be the logical probability function, but this suggestion only succeeds in rendering the theory of information hostage to the difficulties facing the theory of logical probability.

If one is willing to embrace a subjective conception of information, then one might think of taking p to be the subjective probability function of the person whose judgments of informativeness we are explicating. But this has the peculiar result that a person can never be highly confident of any informative proposition. If you acquire evidence that an informative

---

[15] An exception is the notion of "pragmatic information" discussed by Szaniawski (1976).

proposition is probably true, it ceases to be informative. Acquisition of evidence would thus not advance the cognitive goal of accepting true informative propositions.

The weakest, and hence most plausible, interpretation of probabilistic measures of information is Levi's. Levi (1980, p. 48) requires that for each person there be an "information-determining probability" function M, such that the information value of accepting proposition A is measured by the quantity $1 - M(A)$.[16] But he allows the choice of M to be up to the individual, and to be a reflection of the individual's values. Thus the function M is a probability only in the abstract sense that it is required to satisfy the axioms of the probability calculus.

But even Levi's approach rules out certain possible judgments of informativeness – judgments that I think we often actually make. For example, in a partition of n hypotheses, at least one hypothesis must have probability $\leq 1/n$, whatever probability function we use. Thus in any large partition, (9.13) or (9.14) entails that some element of the partition has close to maximal informative content. Yet if the partition concerned the possible outcomes in a long sequence of tosses of a coin, I think we would normally judge that no element of the partition has any great significance. In particular, the outcome of a long sequence of coin tosses is of vastly less significance than, say, the theory of relativistic mechanics; but for a sufficiently long sequence of tosses, some outcome is no more probable than relativity theory, and hence is at least as informative, according to (9.13) or (9.14).

The axioms of probability entail that if p is a probability function, then for all propositions A and B,

$$1 - p(AB) \leq 1 - p(A) + 1 - p(B).$$

So if a measure of informativeness must have the form (9.13), then the information value of a conjunction cannot exceed the

---

[16]Levi allows that a rational person may have unresolved value conflicts, in which case he holds that a rational person must have a *set* of information-determining probabilities, one for each way of resolving the conflicts (Levi 1980, ch. 8). But for a person with no unresolved value conflicts, Levi is committed to the position as I describe it in the text.

235

sum of the information values of its conjuncts; that is, $c(AB) \leq c(A) + c(B)$, for all A and B. But this is not a reasonable condition to impose on a measure of informativeness, as may be seen from the following example. Let A be a proposition concerning the mass of the proton, and let B be a proposition concerning the mass of the neutron. A physicist interested in the structure of atomic nuclei may be very interested in learning how close these masses are to one another, but be much less interested in the absolute magnitudes of the masses. A and B together will tell the physicist how close the masses are to one another, but neither will do so separately. Our physicist therefore values the information provided by A and B together much more than the information either provides separately. So if c measures our physicist's demand for information, we may well have

$$c(AB) > c(A) + c(B).$$

This is a further reason for rejecting the requirement that a measure of information satisfy (9.13), for some probability function p.

A similar objection can be made to the requirement that measures of information satisfy (9.14), for some probability function p. Suppose that in the example of the preceding paragraph, $c(A) = c(B) = 0.1$. Then if c satisfies (9.14) for probability function p, it follows that $p(A) = p(B) = 2^{-0.1}$. Hence

$$
\begin{aligned}
c(AB) &= -\log_2 p(AB) \\
&= -log_2[p(A) + p(B) - p(A \cup B)] \\
&\leq -log_2[2^{-0.1} + 2^{-0.1} - 1] \\
&< 0.21.
\end{aligned}
$$

But in view of the circumstances described, we may well expect $c(AB)$ to be larger than this; for example, there is no reason why our physicist should not have $c(AB) = 0.3$. So (9.14), like (9.13), is not a condition that measures of informational content should be required to satisfy.

I do not deny that some individuals might be committed to having their judgments of informativeness satisfy (9.13) or (9.14), for some probability function p. For persons who have

236

such a commitment, it would be irrational to assess informativeness in a way that violates the relevant condition. What I have argued is that someone who judges informativeness in a different way is not thereby unscientific.

My definition of informational content, Definition 9.2, does not entail that (9.13) or (9.14) is satisfied. The argument I have given in this section shows that this fact is not a defect of my definition, but rather a positive feature.

9.5.3 Distance from truth

The suggestion in Section 6.3 was that science aims at both informativeness and truth. Having just shown how the first of these desiderata may be characterized, I now turn to characterizing the notion of the distance of a proposition from the truth.

The verisimilitude of a proposition is a measure of its closeness to the whole truth. So we could easily define a measure of the distance of a proposition from the whole truth; for example, the distance of A from the whole truth could be measured by $1 - v(A)$. But the notion I wish to explicate here is not distance from the *whole* truth but distance from the *truth*, that is, how far a proposition is from being true.

The distinction can be made formally as follows. If $d(A)$ denotes the distance of A from the truth, in the sense here intended, then it must be the case that

$$d(A) = 0, \quad \text{for all true propositions } A. \qquad (9.15)$$

This condition does not hold for the notion of distance from the whole truth, since a true proposition need not be the whole truth (it need not have a verisimilitude of 1).

The notion of truth itself is objective. It is an objective fact that snow is white, and hence it is an objective fact that it is true that snow is white. But a measure of distance from truth must measure the relative importance of the true and false consequences of a hypothesis; and such measures of relative importance must rest ultimately on our interest in being right on these various topics. So for essentially the same reason that information content is a subjective notion, distance from truth must also be a subjective notion.

The informational content of any given proposition A is the same, whatever the true state may be; but $u(A, x)$ will in general be different in these different states. Thus we can think of distance from truth as the component of cognitive utility that is responsible for the variation of $u(A, x)$ as x is varied. Let $d_x(A)$ be the distance of A from the truth when x is the true state; then a simple definition of the sort just indicated would be

$$d_x(A) = u(A, t) - u(A, x). \qquad (9.16)$$

If (9.16) were adopted as a definition of $d_x(A)$, then a change in scale of the utility function would change the distance of propositions from the truth. This is an undesirable feature, and so we should multiply the right-hand side of (9.16) by a factor that will cancel out changes of scale. Let $u_{\emptyset} = u(\emptyset, x)$, assumed to be a constant independent of x. By (9.7), $u_{\emptyset} < u_T$, and so we can use $u_T - u_{\emptyset}$ as a scaling factor for measurement of distance from truth,[17] giving

$$d_x(A) = \frac{u(A, t) - u(A, x)}{u_T - u_{\emptyset}}.$$

Since $u(\emptyset, t)$ is not defined, this identity defines $d_x(A)$ only for $A \neq \emptyset$. It will be convenient to have $d_x(\emptyset)$ defined also. Hence the definition I will adopt is

Definition 9.3.

$$d_x(A) = \begin{cases} \frac{u(A,t) - u(A,x)}{u_T - u_{\emptyset}} & \text{if } A \neq \emptyset \\ 1 & \text{if } A = \emptyset. \end{cases}$$

I will also use $d(A)$ to denote the actual distance of A from the truth; that is, $d(A) = d_{x^*}(A)$.

---

[17]Other choices of scaling factor could also be used; these would result in a different scale of measurement for distance from truth. But I am avoiding using $u_T - u_X$ here, because (as will become clear shortly) that choice would force the k in u_k to be 1.

For any true proposition A, we now have

$$
\begin{aligned}
d(A) &= d_{x^*}(A) \\
&= \frac{u(A, t) - u(A, x^*)}{u_T - u_\emptyset} \\
&= \frac{u(A, t) - u(A, t)}{u_T - u_\emptyset}, \quad \text{since } x^* \in A \\
&= 0.
\end{aligned}
$$

Thus (9.15) is satisfied by Definition 9.3. By (9.3) and (9.7), we also have that $d(A) \geq 0$, for all A – as it should be.

Definitions 9.1–9.3 together imply that for all $x \in X$ and $A \in \mathbf{X}$,

$$
u(A, x) = (u_T - u_X)c(A) - (u_T - u_\emptyset)d_x(A) + u_X.
$$

(The proof is straightforward.) An equivalent utility function, differing from u only by a positive affine transformation, is

$$
u'(A, x) = \frac{u_T - u_X}{u_T - u_\emptyset} c(A) - d_x(A).
$$

This utility function is identical to the utility function u_k of (9.12), with

$$
k = \frac{u_T - u_X}{u_T - u_\emptyset}.
$$

Hence adoption of Definitions 9.1–9.3 entails that *all* scientific utility functions can be expressed in the form u_k, for some k. Differences in the cognitive utilities of different scientists can thus all be attributed to differences in one or more of:

- The measure of content;
- The measure of distance from truth; or
- The weight put on content, as compared with closeness to truth.

It is also easy to show that if

$$
u(A, x) = kc(A) - d_x(A)
$$

for some functions c and d_x, and constant k, then d_x is a measure of distance from truth, in the sense of Definition 9.3. It follows

239

that the measure of distance from truth, used in the example in Section 6.3, satisfies Definition 9.3.

Definitions 9.1–9.3 entail that

$$v_x(A) = c(A) - \frac{u_T - u_\emptyset}{u_T - u_X} d_x(A).$$

Letting k be as above, this can be written as

$$v_x(A) = c(A) - \frac{1}{k} d_x(A).$$

Thus we have succeeded in factoring verisimilitude into the two components that intuitively are part of it, namely content and closeness to the truth.

9.6 Scientific realism

In the past decade, much of the debate on scientific values has been concerned with van Fraassen's attack on scientific realism. According to van Fraassen (1980, p. 12), scientific realism holds that science aims to accept true theories. Van Fraassen rejected this, and maintained instead a view he called constructive empiricism, which holds that science aims only to accept empirically adequate theories. (A theory is empirically adequate if everything it says about observable phenomena is true.) The account of scientific values outlined in this chapter has implications for that debate about scientific realism – but perhaps not the implications one might expect, as we will now see.

First, we need to be clear about just what is at issue in this debate. To aim at x is not just to value x but also to take steps designed to achieve x. So when van Fraassen denies that science aims at truth, he might be denying that scientists value truth beyond empirical adequacy; or he might merely be denying that they take steps designed to achieve it. For the latter claim to be correct, it would perhaps suffice for scientists to have completely indeterminate probabilities regarding unobservables. So on the latter interpretation, van Fraassen's claims about aims would have no implications for scientific values. But the latter interpretation is not the one van Fraassen intends. He construes the aim of science as the standard of success in science (1989, pp. 189f.), and understands this notion in such a way that it

240

refers to the values put on the different possible outcomes of accepting a scientific theory. Thus van Fraassen is taking realists to hold that in science, acceptance of a true theory is better than one that is false; and he is opposing this by maintaining that it is a matter of indifference whether an accepted theory is correct about unobservables.

So the issue is about scientific values. One might now ask: Is the issue a descriptive one about the values scientists actually have? Or is it a normative one, about the values scientists ought to have? It is neither of these. Van Fraassen is answering the question *What is science?* (1989, p. 189), and so his claims about scientific values are claims about what values a person must have to be a scientist. He is claiming that to be a scientist a person must value empirical adequacy, but need not value correctness about unobservables. This position is consistent with the possibility that many, or even all, scientists do in fact value correctness about unobservables. Van Fraassen's claim is merely that having this value is not essential to being a scientist.[18]

Thus understood, van Fraassen's claim about scientific values is addressed to the same issue that I dealt with in Section 9.2, where I proposed necessary conditions for a person to count as having scientific values. I will now examine the points of similarity and difference between my proposals and van Fraassen's position.

Talk of the aim of an activity, or of a criterion of success of that activity, seems to assume that the activity has only one desirable outcome. The view of scientific values I have been defending in this chapter does not make that assumption. Instead, I have allowed that acceptance of a scientific theory may have a continuum of more or less desirable outcomes; in which case, there is no nonarbitrary dividing line between what counts as "success" and what counts as "failure." For that reason, in this chapter I have chosen to talk of scientific values, represented by a utility function, rather than talking of an "aim" of

[18] The question of whether van Fraassen is making a descriptive or normative claim was posed by Gideon Rosen, in a paper presented at the University of Michigan in April 1990, with van Fraassen responding. Rosen argued that on either of these two interpretations, van Fraassen's position is problematic. Van Fraassen replied to Rosen's paper in the way I have indicated here.

science. This is one point of difference between my proposals and van Fraassen's position.

I do not deny that a scientist *might* have a cognitive utility function in which the acceptance of any theory has only two possible utilities. What I deny is that this is *necessary* for having scientific values. Indeed, I argued that real scientists typically do not satisfy this condition.

I turn now to the claim of van Fraassen's that has provoked the most debate – his claim that science does not aim at truth beyond empirical adequacy. Recall that what this claim means is that someone who does not value truth beyond empirical adequacy is not thereby unscientific. In terms of cognitive utility functions, the claim is this: It is possible for a scientific utility function u to satisfy the condition

$$u(A, t) \;\; = \;\; u(A, x), \quad \text{for all states } x \text{ in which} \\ A \text{ is empirically adequate.} \qquad (9.17)$$

As noted in Section 9.2.1, my conditions on a scientific utility function are consistent with this condition. Hence I agree with van Fraassen that scientists need not value truth beyond empirical adequacy. In fact, there have been very good scientists who have maintained that they cared only for empirical adequacy; so van Fraassen is surely right on this point.

But while a scientific utility function might satisfy (9.17), it need not do so; and I think that most scientists do prefer accepting true theories to ones that are merely empirically adequate. If this is right, then (9.17) does not usually hold for the cognitive utility functions of actual scientists.

That is an empirical claim, and to settle whether or not it is correct would require empirical evidence. This could be done by a survey that would ask scientists for their preferences regarding suitable possible experiments, with the preference interpretation of probability and utility then being used to infer the cognitive utilities we are interested in. (The representation theorem of Chapter 8 shows that this can be done.) I suspect that if such a survey were conducted, we would find that scientists typically have a higher cognitive utility for accepting a theory when it is true than when it is false but empirically adequate.

242

But it would be interesting to have an actual study of this kind conducted.

In any case, this empirical claim is consistent with van Fraassen's constructive empiricism, as I noted earlier.

The positive side of van Fraassen's account of the aim of science is that science aims at empirical adequacy. Since this is a claim about scientific values, it can be expressed in terms of scientific utility. The claim would then presumably be that all scientific utility functions satisfy the condition

$$u(A, x) > u(A, y) \quad \text{whenever } A \text{ is empirically adequate in } x,$$
$$\text{and not empirically adequate in } y. \quad (9.18)$$

While I would allow that scientific utility functions *may* satisfy this condition, I deny that they *must* satisfy it. For example, imagine an individual who accepts theories that make claims about all of space-time, but cares only for whether those theories fit observable phenomena that occur within a billion light years of us. This individual cares about empirical adequacy with regard to all phenomena that our scientists are going to encounter, at least for the next billion years (by which time we probably will be long extinct). I see no compelling reason to say that this individual's values are unscientific. Indeed, I guess that those scientists who say they care only for empirical adequacy, not truth, would also say that they do not care for empirical adequacy in such remote regions of space-time.

So the position on scientific values that I have defended in this chapter is *weaker* than van Fraassen's. It is weaker because it does not require the scientific utility of accepting a theory to have only two possible values, and because it does not require (9.18) to hold. Since van Fraassen's position is in turn weaker than scientific realism (as he defines that view), I could describe my position by saying that I am even less of a scientific realist than van Fraassen is. However, I have suggested that the realist position is likely to be closer to being a correct account of most actual scientists' values.

In any case, neither van Fraassen nor realists dispute the necessary conditions on a scientific utility function that I adopted

243

in Section 9.2. And I have shown that when these conditions are satisfied, then we can say that the cognitive utility of accepting a theory depends only on the content of the theory, and on its distance from the truth. The notions of content and distance from truth are subjective, and different scientific values will be reflected in different measures of content and distance from truth.

Appendix A

Proof for Section 5.1.6

This appendix proves the theorem stated in Section 5.1.6.

Given a suitable set X of states of nature, the expected utility of any act a can be written as[1]

$$EU(a) = \sum_{x \in X} p(x)u(xa)$$

where p and u are the person's probability and utility functions, and xa denotes the conjunction of x and a. According to causal decision theory (which I assume here)[2] X is a suitable set of states for calculating the expected utility of a iff X is a partition such that for each $x \in X$, x is not causally influenced by a, and x determines what consequence will be obtained if a is chosen. Consequences are here understood as including every aspect of the outcome that matters to the person.

Let S be a suitable set of states of nature for calculating the expected utility of the acts that will be available after the choice between d and d' is made. By (iv) the states $s \in S$ are causally independent of d, and hence conjunctions of the form $s.d \to R_q$ are also causally independent of d.[3] Let the set of acts available after choosing d or d' be B, and assume there is a unique $b_q \in B$ that you would choose if your probability function were q. Then d together with $d \to R_q$ determines that you will choose b_q, and this together with s determines the unique consequence you will obtain as a result. Hence conjunctions of the form $s.d \to R_q$ are both causally independent of d and determine what consequence

---

[1] I will assume that the set of states is at most countable. This assumption can be removed by replacing summation with integration.

[2] For a discussion of this theory and its alternative, see e.g., (Maher 1987).

[3] Where ambiguity would otherwise arise, I use a dot to represent conjunction; the scope extends to the end of the formula (or to the next dot, if there is one). So $s.d \to R_q$ denotes the conjunction of s and $d \to R_q$.

245

will eventually be obtained as an indirect result of choosing d. We can therefore take these propositions to be our states for the purpose of computing the expected utility of d. Letting Q be the set of all probability functions that you could have after choosing d, we then have

$$EU(d) = \sum_{s \in S} \sum_{q \in Q} p(s.d \rightarrow R_q) u(sb_q). \qquad (A.1)$$

You know that if d and $d \rightarrow R_q$ are true, then R_q must obtain. Hence $p(R_q | s.d \rightarrow R_q.d) = 1$. Thus (A.1) implies

$$EU(d) = \sum_{s \in S} \sum_{q \in Q} p(s.d \rightarrow R_q) p(R_q | s.d \rightarrow R_q.d) u(sb_q).$$

Applying Bayes' theorem to $p(R_q | s.d \rightarrow R_q.d)$ then gives

$$EU(d) = \sum_{s \in S} \sum_{q \in Q} p(s.d \rightarrow R_q) \frac{p(s.d \rightarrow R_q | R_q d) p(R_q | d)}{p(s.d \rightarrow R_q | d)} u(sb_q).$$

By (v), $p(s.d \rightarrow R_q | d) = p(s.d \rightarrow R_q)$, and so we have

$$EU(d) = \sum_{s \in S} \sum_{q \in Q} p(s.d \rightarrow R_q | R_q d) p(R_q | d) u(sb_q). \qquad (A.2)$$

By condition (a), you are sure one of the $d \rightarrow R_q$ holds, and so $p(s.d \rightarrow R_q | R_q d) = p(s | R_q d)$. I assume that $p(d | R_q) = 1$, for all $q \in Q$; a sufficient condition for this would be that you know what act you have chosen after you choose it. Hence $p(s | R_q d) = p(s | R_q)$, and (A.2) simplifies to

$$
\begin{aligned}
EU(d) &= \sum_{s \in S} \sum_{q \in Q} p(s | R_q) p(R_q | d) u(sb_q) \\
&= \sum_{q \in Q} p(R_q | d) \sum_{s \in S} p(s | R_q) u(sb_q).
\end{aligned}
$$

Since you are sure one of the $d \rightarrow R_q$ is true, $p(R_q | d) = p(d \rightarrow R_q | d)$; and by (v), $p(d \rightarrow R_q | d) = p(d \rightarrow R_q)$. Thus

$$EU(d) = \sum_{q \in Q} p(d \rightarrow R_q) \sum_{s \in S} p(s | R_q) u(sb_q). \qquad (A.3)$$

246

Condition (b) asserts that for each $q \in Q$ there is a probability function q' such that $p(d' \to R_{q'}|d \to R_q) = 1$. Obviously, only one q' can satisfy this condition for a given q; I will use the notation $\varphi(q)$ to denote this q'. I will also use Q' to denote the set of probability functions you might have after choosing d'. Then we can rearrange the summation over Q in (A.3), to give

$$EU(d) = \sum_{q' \in Q'} \sum_{\varphi(q)=q'} p(d \to R_q) \sum_{s \in S} p(s|R_q)u(sb_q).$$

By condition (d), $p(s|R_q) = p(s|R_{\varphi(q)})$. Also, condition (i) entails that $p(s|R_{\varphi(q)}) = \varphi(q)(s)$. Hence

$$
\begin{aligned}
EU(d) &= \sum_{q' \in Q'} \sum_{\varphi(q)=q'} p(d \to R_q) \sum_{s \in S} \varphi(q)(s)u(sb_q) \\
&\leq \sum_{q' \in Q'} \sum_{\varphi(q)=q'} p(d \to R_q) \max_{b \in B} \sum_{s \in S} \varphi(q)(s)u(sb) \\
&= \sum_{q' \in Q'} \left[\sum_{\varphi(q)=q'} p(d \to R_q) \right] \max_{b \in B} \sum_{s \in S} q'(s)u(sb). \quad \text{(A.4)}
\end{aligned}
$$

By the definition of φ, $p(d' \to R_{q'}|d \to R_q)$ equals 1 if $q' = \varphi(q)$, and 0 otherwise. Hence

$$\sum_{\varphi(q)=q'} p(d \to R_q) = \sum_{q \in Q} p(d' \to R_{q'}|d \to R_q)p(d \to R_q).$$

The theorem of total probability then gives

$$\sum_{\varphi(q)=q'} p(d \to R_q) = p(d' \to R_{q'}).$$

Substituting in (A.4), we have

$$EU(d) \leq \sum_{q' \in Q'} p(d' \to R_{q'}) \max_{b \in B} \sum_{s \in S} q'(s)u(sb). \quad \text{(A.5)}$$

The reasoning leading to (A.3) can be repeated, mutatis mutandis, for d', giving

$$EU(d') = \sum_{q' \in Q'} p(d' \to R_{q'}) \sum_{s \in S} p(s|R_{q'})u(sb_{q'}),$$

247

From condition (i) we have $p(s|R_{q'}) = q'(s)$, and so

$$
\begin{aligned}
EU(d') &= \sum_{q' \in Q'} p(d' \to R_{q'}) \sum_{s \in S} q'(s) u(sb_{q'}) \\
&= \sum_{q' \in Q'} p(d' \to R_{q'}) \max_{b \in B} \sum_{s \in S} q'(s) u(sb) \quad \text{(A.6)}
\end{aligned}
$$

Comparing (A.5) and (A.6), we see that $EU(d) \leq EU(d')$.

Appendix B

Proof of Theorem 8.1

This appendix proves the representation theorem for simple cognitive expected utility stated in Section 8.2. Throughout this appendix, it is assumed that Axioms 1–8 are satisfied.

B.1 PROBABILITY

The following theorem establishes the existence of a quantitative subjective probability function that agrees with the qualitative probability relation discussed in Section 8.2.8.

Theorem B.1. *Let* $\precsim \bullet$ *be a binary relation defined on* **X** *by the condition that for all* $F, G \in \mathbf{X}$, $F \precsim \bullet G$ *iff there exist* $a, b \in Y$ *and* $f, g \in \mathbf{D}$ *such that*

- $a \prec b;$
- $f = b$ *on* F *and* a *on* $\bar{F}$;
- $g = b$ *on* G *and* a *on* $\bar{G}$; *and*
- $f \precsim g$.

Then there exists a unique function p *such that for all* $F, G \in \mathbf{X}$:

(i) $\langle X, \mathbf{X}, p \rangle$ *is a (finitely additive) probability space.*

(ii) $p(F) \leq p(G)$ *iff* $F \precsim \bullet G$.

(iii) $p(F) = 0$ *iff* $F \in \mathbf{N}$.

(iv) If $\rho \in [0, 1]$ *there exists* $A \in \mathbf{X}$ *such that* $A \subset F$ *and* $p(A) = \rho p(G)$.

Savage (1954, ch. 3) shows that Theorem B.1 follows from his postulates. His proof remains valid when the axioms of Section 8.1 are used in place of his postulates; consequently, the reader is referred to Savage for a proof of this theorem.

Throughout this appendix, 'p' denotes the unique probability function asserted to exist by Theorem B.1.

In the next section, we will make use of the following theorem.

249

Theorem B.2. *Let* $\{A_1, \ldots, A_n\} \subset \mathbf{X}$ *be a partition of* $A \in \mathbf{X}$, *and let* $f, g \in \mathbf{D}$. *If* $f \precsim g$ *given* A_i, *for all* $i = 1, \ldots, n$, *then* $f \precsim g$ *given* A. *And if in addition there exists* i *such that* $f \prec g$ *given* A_i, *then* $f \prec g$ *given* A.

As Savage (1954, p. 24) remarks, this can be proved using mathematical induction.

B.2 UTILITY OF GAMBLES

Let $a_1, \ldots, a_n \in Y$ and let $\rho_1, \ldots, \rho_n \geq 0$, where $\sum_{i=1}^n \rho_i = 1$. I will use $\sum_{i=1}^n \rho_i a_i$ to denote the set of all $f \in \mathbf{D}$ such that for some partition $\{A_1, \ldots, A_n\}$ of X, $f_i = a_i$ on A_i, and $p(A_i) = \rho_i$, $i = 1, \ldots, n$. A subset F of $\mathbf{D}$ is said to be a *gamble* (Savage 1954) or *lottery* (Fishburn 1981) if $F \neq \emptyset$ and

$$F = \sum_{i=1}^n \rho_i a_i$$

for some a_i and ρ_i. The set of all gambles will be denoted by $\mathbf{L}$.

The purpose of this section is to prove the existence of a utility function on $\mathbf{L}$. But first we need a well-defined notion of preference on $\mathbf{L}$. The following theorem is addressed to that need, showing that gambles are equivalence classes in $\bigcup \mathbf{L}$,[1] whence $\precsim$ on $\mathbf{D}$ induces a corresponding ordering of $\mathbf{L}$. (In this theorem, and subsequently, the abbreviation 'a.e.' means 'almost everywhere', i.e., on a set of probability 1.)

Theorem B.3. *Let* $\sum_{i=1}^n \rho_i a_i \in \mathbf{L}$, *let* $f', g' \in \sum_{i=1}^n \rho_i a_i$, *and let* $f, g \in \mathbf{D}$ *be such that* $f = f'$ *a.e. and* $g = g'$ *a.e.. Then* $f \sim g$.

Proof. It is easily shown, using Theorem B.1 and Definition 8.2, that $f \sim f'$ and $g \sim g'$. Thus there is no loss of generality in supposing that $f, g \in \sum_{i=1}^n \rho_i a_i$.

I will first prove that the theorem holds for $n = 2$, and afterward show that the general case follows.

Suppose then that $f, g \in \rho_1 a_1 + \rho_2 a_2$. We may assume, without loss of generality, that $f \precsim g$ given $f^{-1}(a_1) \cap g^{-1}(a_2)$, that

---

[1] $\bigcup \mathbf{L}$ is the union of the elements of $\mathbf{L}$. It is identical to the set of simple acts.

is, that one or the other of the following three cases holds:

(i) $f^{-1}(a_1) \cap g^{-1}(a_2) \in \mathbf{N}$.

(ii) $f \sim g$ given $f^{-1}(a_1) \cap g^{-1}(a_2)$, and
$f^{-1}(a_1) \cap g^{-1}(a_2) \notin \mathbf{N}$.

(iii) $f \prec g$ given $f^{-1}(a_1) \cap g^{-1}(a_2)$.

If (i), then since (as is easily verified)

$$p\left[f^{-1}(a_1) \cap g^{-1}(a_2)\right] = p\left[f^{-1}(a_2) \cap g^{-1}(a_1)\right]$$

we have by Theorem B.1 that $f^{-1}(a_2) \cap g^{-1}(a_1) \in \mathbf{N}$, whence $f \sim g$ given $f^{-1}(a_2) \cap g^{-1}(a_1)$. If (ii), then by Axiom 6, $f \sim g$ given $f^{-1}(a_2) \cap g^{-1}(a_1)$. So whether (i) or (ii) holds, we have by Theorem B.2 that $f \sim g$. Turning now to case (iii), let $b_1, b_2 \in Y$ be such that $b_1 \prec b_2$, and define $f', g' \in \mathbf{D}$ by

$$f' = \left\{ \begin{array}{l} b_1 \text{ on } f^{-1}(a_1) \\ b_2 \text{ on } f^{-1}(a_2) \end{array} \right\}, \quad g' = \left\{ \begin{array}{l} b_1 \text{ on } g^{-1}(a_1) \\ b_2 \text{ on } g^{-1}(a_2) \end{array} \right\}.$$

Then $f' \prec g'$ given $f^{-1}(a_1) \cap g^{-1}(a_2)$, by Axiom 6. So by Axiom 7, $f \precsim g$ iff $f' \precsim g'$. But by Theorem B.2 and the fact that $p[f^{-1}(a_1)] = p[g^{-1}(a_1)]$, $f' \sim g'$. Hence $f \sim g$. This completes the proof for $n = 2$.

Turning now to the general case, let $C \in \mathbf{X}$ be such that for $i, j = 1, \ldots, n$

$$p\left[C \cap f^{-1}(a_i) \cap g^{-1}(a_j)\right] = p\left[f^{-1}(a_i) \cap g^{-1}(a_j)\right]/2.$$

(The existence of such a C is guaranteed by Theorem B.2.) Let $a \in Y$ be such that $h = a$ for some $h \in \mathbf{D}$, and define $f_1, f_2, g_1 \in \mathbf{D}$ by

$$f_1 = \left\{ \begin{array}{l} f \text{ on } C \\ a \text{ on } \bar{C} \end{array} \right\}, \quad f_2 = \left\{ \begin{array}{l} a \text{ on } C \\ f \text{ on } \bar{C} \end{array} \right\}, \quad g_1 = \left\{ \begin{array}{l} g \text{ on } C \\ a \text{ on } \bar{C} \end{array} \right\}.$$

By the $n = 2$ case already proved, we have that for all $i = 1, \ldots, n$, $f_1 \sim f_2$ given $f^{-1}(a_i)$. So by Theorem B.2, $f_1 \sim f_2$. Furthermore, since

$$p\left[f^{-1}(a_i) \cap \bar{C}\right] = p\left[g^{-1}(a_i) \cap C\right],$$

251

Theorem B.1 implies that $f_2 \sim g_1$ given $[f^{-1}(a_i) \cap \bar{C}] \cup [g^{-1}(a_i) \cap C]$, $i = 1, \ldots, n$. But

$$\left\{ \left[f^{-1}(a_i) \cap \bar{C} \right] \cup \left[g^{-1}(a_i) \cap C \right] : i = 1, \ldots, n \right\}$$

is a partition of X, as so we have by Theorem B.2 that $f_2 \sim g_1$. So by the transitivity of $\sim$, $f_1 \sim g_1$. Consequently, $f \sim g$ given C. Interchanging the roles of C and $\bar{C}$ in this proof gives also that $f \sim g$ given $\bar{C}$, and so $f \sim g$. ∎

(The solid block is used to mark the end of a proof.)

The preference relation $\precsim$ can now be extended to **L** in the obvious way.

Definition B.1. *Let $F, G \in \mathbf{L}$, $f \in F$, $g \in G$. Then $F \precsim G$ iff $f \precsim g$.*

The relations $\prec$ and $\sim$ on **L** are of course defined in terms of $\precsim$ in the usual way.

I will now define the notion of a *mixture* of gambles. Let $F_1, \ldots, F_n \in \mathbf{L}$, and for $i = 1, \ldots, n$ let $F_i = \sum_{j=1}^{m} \rho_{ij} a_j$. (Note that this latter condition involves no loss of generality, since some of the ρ_{ij} may be zero.) Then if $\sigma_1, \ldots, \sigma_n \geq 0$, and $\sum_{i=1}^{n} \sigma_i = 1$, the mixture

$$\sum_{i=1}^{n} \sigma_i F_i$$

of $F_1, \ldots, F_n$ is defined to be

$$\sum_{i=1}^{n} \sum_{j=1}^{m} (\sigma_i \rho_{ij}) a_j.$$

The next theorem shows that **L** is closed under this mixing operation. In other words, **L** is a convex set.

Theorem B.4. *Let $F_1, \ldots, F_n \in \mathbf{L}$, and let $\sigma_1, \ldots, \sigma_n \geq 0$, with $\sum_{i=1}^{n} \sigma_i = 1$. Then $\sum_{i=1}^{n} \sigma_i F_i \in \mathbf{L}$.*

252

Proof. What needs to be shown is that $\sum_{i=1}^{n} \sigma_i F_i$ is nonempty. This will be proved by induction on n.

If $n = 1$, then $\sum_{i=1}^{n} \sigma_i F_i = F_1$, and then since F_1 is nonempty by definition, so is $\sum_{i=1}^{n} \sigma_i F_i$.

Suppose now that the theorem is true for some n, and consider the mixture $\sum_{i=1}^{n+1} \sigma_i F_i$ of gambles $F_1, \ldots, F_{n+1}$. If $\sigma_{n+1} = 1$, then $\sum_{i=1}^{n+1} \sigma_i F_i = F_{n+1}$, which is nonempty. Suppose then that $\sigma_{n+1} < 1$, and let

$$F = \sum_{i=1}^{n} \frac{\sigma_i}{1 - \sigma_{n+1}} F_i.$$

By assumption, there exists $f \in F$ and $g \in F_{n+1}$. By Theorem B.1 we have that for each $a, b \in Y$ there exists $A_{a,b} \in \mathbf{X}$ such that

$$p\left[A_{a,b} \cap f^{-1}(a) \cap g^{-1}(b) \right] = \sigma_{n+1} p\left[f^{-1}(a) \cap g^{-1}(b) \right].$$

Let

$$A = \bigcup_{a,b \in Y} A_{a,b}.$$

Since only finitely many $A_{a,b}$ are nonempty, $A \in \mathbf{X}$. Hence by Axiom 3 there exists $h \in \mathbf{D}$ such that $h = f$ on $\bar{A}$ and g on A. Then

$$h \in (1 - \sigma_{n+1})F + \sigma_{n+1} F_{n+1} = \sum_{i=1}^{n+1} \sigma_i F_i.$$

Thus the theorem holds with $n + 1$ in place of n. ∎

We are now in a position to define the notion of a utility on **L**.

Definition B.2. *A real-valued function v on* **L** *is a* utility on **L** *iff for all $F, G \in \mathbf{L}$ and $\rho \in [0, 1]$,*

(i) $F \precsim G$ iff $v(F) \le v(G)$.
(ii) $v(\rho F + (1 - \rho)G) = \rho v(F) + (1 - \rho)v(G)$.

In other words, a utility on **L** is an order-preserving linear real-valued function on **L**.

The next task is to prove that there is a utility on **L**. Not to belabor familiar material, I will state without proof the following theorem.

253

Theorem B.5. *There exists a utility on* **L** *iff for all* $F, G, H \in$ **L** *the following three conditions hold:*

(i) $\precsim$ *is connected and transitive on* **L**.

(ii) *If* $\rho \in (0, 1)$, *then* $F \precsim G$ *iff*
$$\rho F + (1 - \rho)H \precsim \rho G + (1 - \rho)H.$$

(iii) *If* $F \prec G \prec H$, *then there exist* $\rho, \sigma \in (0, 1)$ *such that*
$$\rho F + (1 - \rho)H \prec G \prec \sigma F + (1 - \sigma)H.$$

Essentially Theorem B.5 is due to von Neumann and Morgenstern (1947), but for a proof in which the assumptions have the form (i)–(iii) see (Jensen 1967).

It is an immediate consequence of Definition B.1 and Axioms 1 and 4 that condition (i) of Theorem B.5 is satisfied. Thus the existence of a utility on **L** will be proved if we prove that conditions (ii) and (iii) are also satisfied. To this we now turn.

Theorem B.6. *Let* $F, G \in$ **L** *with* $f \in F$ *and* $g \in G$, *and suppose* $A \in \mathbf{X} \setminus \mathbf{N}$ *is independent of* $f^{-1}(a) \cap g^{-1}(b)$, *for all* $a, b \in Y$. *Then* $f \precsim g$ *iff* $f \precsim g$ *given* A.

Proof. I will first show that, if $B \in \mathbf{X}$ is independent of $f^{-1}(a) \cap g^{-1}(b)$ for all $a, b \in Y$, and if $p(A) = p(B)$, then $f \precsim g$ given A iff $f \precsim g$ given B. To this end, let $c \in Y$ be such that $h = c$ for some $h \in \mathbf{D}$, and define $f', f'', g', g'' \in \mathbf{D}$ by the conditions

$$f' = \left\{ \begin{array}{l} f \text{ on } A \\ c \text{ on } \bar{A} \end{array} \right\}, \qquad g' = \left\{ \begin{array}{l} g \text{ on } A \\ c \text{ on } \bar{A} \end{array} \right\},$$

$$f'' = \left\{ \begin{array}{l} f \text{ on } B \\ c \text{ on } \bar{B} \end{array} \right\}, \qquad g'' = \left\{ \begin{array}{l} g \text{ on } B \\ c \text{ on } \bar{B} \end{array} \right\}.$$

Then letting ρ be the common value of $p(A)$ and $p(B)$, we have that $f', f'' \in \rho F + (1 - \rho)c$ and $g', g'' \in \rho G + (1 - \rho)c$. So by Theorem B.3, $f' \sim f''$ and $g' \sim g''$. Thus $f' \precsim g'$ iff $f'' \precsim g''$, and from this we obtain the desired result, viz. $f \precsim g$ given A iff $f \precsim g$ given B.

In view of the result just established, it will be convenient to introduce the (temporary) notation '$f \precsim g$ given ρ', where

254

$\rho \in [0,1]$. This notation will mean that there exists $B \in \mathbf{X}$ such that B is independent of $f^{-1}(a) \cap g^{-1}(b)$ for all $a, b \in Y$, $p(B) = \rho$, and $f \precsim g$ given B.

Next the following three claims will be established. (Here and subsequently, I^+ denotes the set of positive integers.)

(i) If $\rho \in (0,1]$, $n \in I^+$, and $n \le \rho^{-1}$, then $f \precsim g$ given ρ only if $f \precsim g$ given $n\rho$.

(ii) If $\rho \in [0,1]$ and $n \in I^+$, then $f \precsim g$ given ρ only if $f \precsim g$ given $n^{-1}\rho$.

(iii) If $\rho \in (0,1]$ and $f \prec g$ given ρ, then there exists $\varepsilon > 0$ such that $f \prec g$ given σ, for all $\sigma \in [\rho - \varepsilon, \rho]$.

To prove (i), note that by Theorem B.1 there exist pairwise disjoint $A_1, \ldots, A_n \in \mathbf{X}$ such that for all $i = 1, \ldots, n$ and $a, b \in Y$,

$$p\left[A_i \cap f^{-1}(a) \cap g^{-1}(b)\right] = \rho p \left[f^{-1}(a) \cap g^{-1}(b)\right].$$

By assumption, $f \precsim g$ given A_i, $i = 1, \ldots, n$. Hence by Theorem B.2, $f \precsim g$ given $\bigcup_{i=1}^{n} A_i$. Since $p\left[\bigcup_{i=1}^{n} A_i\right] = n\rho$, this is the desired result.

For the proof of (ii), suppose there exists $n \in I^+$ such that $g \prec f$ given $n^{-1}\rho$. Then one can show, much as in the proof of (i), that $g \prec f$ given ρ. So by contraposition, $f \precsim g$ given ρ only if $f \precsim g$ given $n^{-1}\rho$.

For the proof of (iii), let $B \in \mathbf{X}$ be such that for all $a, b \in Y$,

$$p\left[B \cap f^{-1}(a) \cap g^{-1}(b)\right] = \rho p \left[f^{-1}(a) \cap g^{-1}(b)\right].$$

Define $f_0 \in \mathbf{D}$ to be the act that agrees with f on B, and g on $\bar{B}$. Then by assumption, $f_0 \prec g$. Now let $A_1, \ldots, A_n$ be an enumeration of

$$\left\{f^{-1}(a) \cap g^{-1}(b) : a, b \in Y\right\} \setminus \mathbf{N}$$

such that if $f \precsim g$ given A_i and $g \prec f$ given A_j, then $i < j$. We define simultaneously a sequence $f_1, \ldots, f_n$ in $\mathbf{D}$ and a sequence

255

$B_1, \ldots, B_n$ in $\mathbf{X}$ by the following two conditions, held true for each $i \in \{1, \ldots, n\}$:

(a) Let $C_1, \ldots, C_n$ be a partition of A_i such that, for all $j = 1, \ldots, m$ if

$$h = \left\{ \begin{array}{l} g \text{ on } C_j \\ f_{i-1} \text{ on } \bar{C}_j \end{array} \right\}$$

then $h \prec g$. Choose some j such that $C_j \cap B \notin \mathbf{N}$, and let $B_i = C_j \cap B$.

(b) Let

$$f_i = \left\{ \begin{array}{l} g \text{ on } B_i \\ f_{i-1} \text{ on } \bar{B}_i \end{array} \right\}.$$

(Note that $f_i \prec g$ for all $i = 1, \ldots, n$.) Now define

$$\varepsilon = \min \left\{ p(B_i)/p(A_i) : i = 1, \ldots, n \right\}.$$

For each $i = 1, \ldots, n$, let $D_i \in \mathbf{X}$ be such that $D_i \subset B_i$ and $p(D_i) = \varepsilon p(A_i)$. Let $D = \bigcup_{i=1}^n D_i$, and define $f' \in \mathbf{D}$ by the condition

$$f' = \left\{ \begin{array}{l} f \text{ on } B \setminus D \\ g \text{ elsewhere} \end{array} \right\}.$$

Then if $m = \max\{i : f \precsim g \text{ given } A_i\}$, we have by Theorem B.2 that $f' \prec f_m \prec g$. And so, since

$$p[A_i \cap (B \setminus D)] = (\rho - \varepsilon)p(A_i), \quad i = 1, \ldots, n$$

we have that $f \prec g$ given $\rho - \varepsilon$. Now if $0 < \varepsilon' < \varepsilon$, substituting ε' for ε in the proof just given shows that $f \prec g$ given $\rho - \varepsilon'$. This completes the proof of (iii).

With (i)–(iii) established, I proceed to complete the proof of the theorem. For this it suffices to show that for all $\rho \in (0, 1]$, $f \precsim g$ given ρ iff $f \precsim g$. That result is trivial if, for all $\rho \in (0, 1]$, $f \sim g$ given ρ. So suppose that for some $\sigma \in (0, 1]$, $f \not\sim g$ given σ. Without loss of generality, let $f \prec g$ given σ. By (iii) there exists $\varepsilon > 0$ such that $f \prec g$ given τ, for all $\tau \in [\sigma - \varepsilon, \sigma]$. Now suppose that for some $\rho \in (0, 1]$,

$$g \precsim f \text{ given } \rho. \tag{B.1}$$

Choose $m, n \in I^+$ such that $n \geq m\rho$ and $mn^{-1}\rho \in [\sigma - \varepsilon, \sigma]$. Then (ii) gives that

$$g \prec f \text{ given } n^{-1}\rho$$

and hence by (i) it follows that

$$g \precsim f \text{ given } mn^{-1}\rho.$$

But this contradicts the choice of m and n, and hence (B.1) is false. Thus we have shown that $f \prec g$ given ρ, for all $\rho \in (0, 1]$, and hence that $f \precsim g$ given ρ iff $f \precsim g$. ∎

The following theorem asserts that condition (ii) of Theorem B.5 is satisfied. It is an easy corollary of Theorem B.6.

Theorem B.7. *If $F, G, H \in \mathbf{L}$, and $\rho \in (0, 1]$, then*

$$\rho F + (1 - \rho)H \precsim \rho G + (1 - \rho)H \quad \text{iff} \quad F \precsim G.$$

Proof. Let $f \in F$, $g \in G$, and $h \in H$. Choose $A \in \mathbf{X}$ such that for all $a, b, c \in Y$,

$$p\left[A \cap f^{-1}(a) \cap g^{-1}(b) \cap h^{-1}(c)\right]$$
$$= \rho p\left[f^{-1}(a) \cap g^{-1}(b) \cap h^{-1}(c)\right].$$

Define $f', g' \in \mathbf{D}$ by the conditions

$$f' = \left\{ \begin{array}{l} f \text{ on } A \\ h \text{ on } \bar{A} \end{array} \right\}, \qquad g' = \left\{ \begin{array}{l} g \text{ on } A \\ h \text{ on } \bar{A} \end{array} \right\}.$$

Then $f' \in \rho F + (1-\rho)H$, and $g' \in \rho G + (1-\rho)H$. Thus we have the following chain of equivalences (the third of which holds by Theorem B.6):

257

$$\rho F + (1 - \rho)H \precsim \rho G + (1 - \rho)H \iff f' \precsim g'$$
$$\iff f \precsim g \text{ given } A$$
$$\iff f \precsim g$$
$$\iff F \precsim G. \quad \blacksquare$$

The next theorem asserts that condition (iii) of Theorem B.5 is also satisfied.

Theorem B.8. *If* $F, G, H \in \mathbf{L}$, *and* $F \prec G \prec H$, *then there exist* $\rho, \sigma \in (0, 1)$ *such that*

$$\rho F + (1 - \rho)H \prec G \prec \sigma F + (1 - \sigma)H.$$

Proof. Let $f \in F$, $g \in G$, $h \in H$, and let $A_1, \ldots, A_n$ be an enumeration of

$$\left\{ f^{-1}(a) \cap h^{-1}(b) : a, b \in Y \right\} \setminus \mathbf{N}$$

such that if $f \precsim h$ given A_i and $h \prec f$ given A_j, then $i < j$. By Axiom 8, there exists a sequence $f_0, f_1, \ldots, f_n$ in $\mathbf{D}$ and a sequence $B_1, \ldots, B_n$ in $\mathbf{X} \setminus \mathbf{N}$ such that for all $i = 1, \ldots, n$:

(i) $f_0 = f$

(ii) $B_i \subset A_i$

(iii) $f = \left\{ \begin{array}{l} h \text{ on } B_i \\ f_{i-1} \text{ on } \bar{B}_i \end{array} \right\}$

(iv) $f_i \prec g$.

Let

$$\varepsilon = \min \left\{ p(B_i)/p(A_i) : i = 1, \ldots, n \right\}$$

and for each $i = 1, \ldots, n$ let $C_i \in \mathbf{X}$ be such that $C_i \subset B_i$ and $p(C_i) = \varepsilon p(A_i)$. Let $B = \bigcup_{i=1}^{n} B_i$ and $C = \bigcup_{i=1}^{n} C_i$, and define $f' \in \mathbf{D}$ by

$$f' = \left\{ \begin{array}{l} h \text{ on } C \\ f \text{ on } \bar{C} \end{array} \right\}.$$

Then if $m = \max\{i : f \precsim h \text{ given } A_i\}$, we have by Theorem B.2

258

that $f' \prec f_m \prec g$. But $f' \in (1 - \varepsilon)F + \varepsilon H$, so setting $\rho = 1 - \varepsilon$ we have that $\rho \in (0, 1)$ and

$$\rho F + (1 - \rho)H \prec G.$$

A similar proof establishes the dual result, namely that there exists $\sigma \in (0, 1)$ such that

$$G \prec \sigma F + (1 - \sigma)H. \quad \blacksquare$$

We have now shown that all conditions (i)–(iii) in Theorem B.5 hold, and hence have established

Theorem B.9. *There exists a utility on* **L**.

It is a familiar consequence of Definition B.2 that this utility on **L** is unique up to a positive affine transformation; that is, we have

Theorem B.10. *If v is a utility on* **L**, *then a real-valued function v' on* **L** *is a utility on* **L** *iff there exist real numbers ρ and σ, $\rho > 0$, such that $v' = \rho v + \sigma$.*

A proof of Theorem B.10 may be found in many places, including the works cited in connection with Theorem B.5.

B.3 UTILITY OF CONSEQUENCES

Let Y^* be the set of all $a \in Y$ such that, for some $f \in$ **D**, $f^{-1}(a) \notin$ **N**. The notion of utility of consequences I am about to introduce will be called a *utility on Y^**, even though the domain of this utility function is the whole of Y. The reason for this terminology is that the values assumed by this utility function on $Y \setminus Y^*$ are quite arbitrary. The term *utility on Y* will therefore be reserved for the stricter notion of utility of consequences to be introduced in Appendix D. The definition of the present notion of utility of consequences is as follows (where $\mathcal{E}(Z)$ is the expected value of the random variable Z):

Definition B.3. *A real-valued function u on Y is a utility on Y^* iff for all $f, g \in \bigcup$ **L**, $f \precsim g$ iff $\mathcal{E}[u(f)] \leq \mathcal{E}[u(g)]$.*

259

In other words, a utility on Y^* is a function whose expected value $\mathcal{E}[u(\cdot)]$ is order preserving on $\bigcup \mathbf{L}$. In fact, $\mathcal{E}[u(\cdot)]$ is order preserving on acts that are almost equal to acts in $\bigcup \mathbf{L}$. More precisely, if $f', g' \in \bigcup \mathbf{L}$, and $f, g \in \mathbf{D}$, and if $f = f'$ a.e. and $g = g'$ a.e., then $f \precsim g$ iff $\mathcal{E}[u(f)] \leq \mathcal{E}[u(g)]$. (That this is so is an immediate consequence of Definition B.3 and clause (iii) of Theorem B.1.)

The aim of the present section is to establish the existence of a utility on Y^*. But first I will establish a uniqueness condition. This asserts that utilities on Y^* are unique on Y^* up to a positive affine transformation, although (as remarked above) there are no restrictions on the values they may take on $Y \setminus Y^*$. Formally, the condition is

Theorem B.11. *Let u be a utility on Y^*. Then a real-valued function u' on Y is a utility on Y^* iff there exist real numbers ρ and σ, $\rho > 0$, such that $u_* = \rho u'_* + \sigma$, where u_* and u'_* are the restrictions of u and u' respectively to Y^*.*

Proof. Let $f, g \in \bigcup \mathbf{L}$, and suppose first that for some $\rho > 0$ and real number σ, $u_* = \rho u'_* + \sigma$. Since $p\left[f^{-1}(Y \setminus Y^*)\right] = p\left[g^{-1}(Y \setminus Y^*)\right] = 0$, we have

$$\mathcal{E}[u(f)] = \mathcal{E}[\rho u'(f) + \sigma] = \rho \mathcal{E}[u'(f)] + \sigma$$

and similarly for g. Applying first Definition B.3 and then the preceding identity gives

$$f \precsim g \iff \mathcal{E}[u(f)] \leq \mathcal{E}[u(g)]$$
$$\iff \mathcal{E}[u'(f)] \leq \mathcal{E}[u'(g)].$$

So u' is a utility on Y^*.

For the other half of the proof, suppose u' is a utility on Y^*. Let $a, b \in Y$, $a \prec b$. If we set

$$\rho = \frac{u(b) - u(a)}{u'(b) - u'(a)}$$

then $\rho > 0$. Then setting $\sigma = u(a) - \rho u'(a)$ gives

$$\rho u'(a) + \sigma = u(a)$$
$$\rho u'(b) + \sigma = u(b).$$

260

Let $u'' = \rho u' + \sigma$. Then by the half of the theorem already proved, u'' is a utility on Y^*. The proof will be completed by showing that for any $c \in Y^*$, $u''(c) = u(c)$.

Since $c \in Y^*$, there exists $h \in \mathbf{D}$ such that $h^{-1}(c) \notin \mathbf{N}$. Define $k \in \bigcup \mathbf{L}$ by

$$k = \left\{ \begin{array}{l} c \text{ on } h^{-1}(c) \\ a \text{ elsewhere} \end{array} \right\}.$$

We consider three cases.

First, suppose $a \precsim k \precsim b$. Now it is a consequence of conditions (i)–(iii) of Theorem B.5 that for any $F, G, H \in \mathbf{L}$, if $F \precsim G \precsim H$ then there exists $\tau \in [0, 1]$ such that $G \sim \tau F + (1 - \tau)H$. (For a proof of this claim, see e.g., [Jensen 1967].) Hence there exists $\tau \in [0, 1]$ such that

$$k \sim \tau a + (1 - \tau)b.$$

So since u'' is a utility on Y^*,

$$\mathcal{E}[u''(k)] = \mathcal{E}[u''(\tau a + (1 - \tau)b].$$

So if $\eta = p\left[k^{-1}(c)\right]$, we have

$$\eta u''(c) + (1 - \eta)u''(a) = \tau u''(a) + (1 - \tau)u''(b). \qquad \text{(B.2)}$$

Similarly, the fact that u is a utility on Y^* gives

$$\eta u(c) + (1 - \eta)u(a) = \tau u(a) + (1 - \tau)u(b). \qquad \text{(B.3)}$$

But (B.2) and (B.3), together with the facts that $u''(a) = u(a)$, $u''(b) = u(b)$, and $\eta > 0$ imply $u''(c) = u(c)$.

For the second case, suppose $k \prec a$. Then by the result appealed to in the preceding paragraph, there exists $\tau \in (0, 1)$ such that

$$a \sim \tau[\eta c + (1 - \eta)a] + (1 - \tau)b.$$

So since u'' is a utility on Y^*,

$$\mathcal{E}[u''(a)] = \mathcal{E}[u''[\tau \eta c + \tau(1 - \eta)a + (1 - \tau)b]].$$

261

That is,

$$u''(a) = \tau \eta u''(c) + \tau(1 - \eta)u''(a) + (1 - \tau)u''(b). \qquad \text{(B.4)}$$

Similarly, the fact that u is a utility on Y^* gives

$$u(a) = \tau \eta u(c) + \tau(1 - \eta)u(a) + (1 - \tau)u(b). \qquad \text{(B.5)}$$

It follows from (B.4) and (B.5) again that $u''(c) = u(c)$.

Finally, if $b \prec k$, then a similar argument shows that in this case too $u''(c) = u(c)$. ∎

Theorem B.12. *There exists a utility on Y^*.*

Proof. Let $a \in Y$ be such that for some $f \in D$, $f = a$. Let $b \in Y^*$, and let $\rho\sigma \in (0, 1]$ be such that $\rho b + (1 - \rho)a \in \mathbf{L}$ and $\sigma b + (1 - \sigma)a \in \mathbf{L}$. Supposing without loss of generality that $\rho \leq \sigma$, we have

$$\rho b + (1 - \rho)a = \frac{\rho}{\sigma}[\sigma b + (1 - \sigma)a] + \frac{\sigma - \rho}{\sigma}a.$$

Thus if v is a utility on $\mathbf{L}$ such that $v(a) = 0$, we have by Definition B.2

$$v[\rho b + (1 - \rho)a] = \frac{\rho}{\sigma}v[\sigma b + (1 - \sigma)a].$$

That is,

$$v[\rho b + (1 - \rho)a]/\rho = v[\sigma b + (1 - \sigma)a]/\sigma.$$

In view of this result, we can define a function u on Y as follows: For any $b \in Y$,

$$u(b) = \left\{ \begin{array}{ll} v[\rho b + (1 - \rho)a]/\rho & \text{if } b \in Y^* \\ 0 & \text{if } b \in Y \setminus Y^* \end{array} \right\}$$

where $\rho > 0$ is such that (if $b \in Y^*$ then) $\rho b + (1 - \rho)a \in \mathbf{L}$. I will now show that the function u thus defined is a utility on Y^*.

262

It suffices to show that for all $\rho_1, \ldots, \rho_m \in [0,1]$ and $a_1, \ldots, a_m \in Y$, if $\sum_{i=1}^{m} \rho_i a_i \in L$, then

$$v\left(\sum_{i=1}^{m} \rho_i a_i\right) = \sum_{i=1}^{m} \rho_i u(a_i). \tag{B.6}$$

For if $f, g \in \bigcup L$, then $f \in \sum_{i=1}^{m} \rho_i a_i$ and $g \in \sum_{j=1}^{n} \sigma_j b_j$, for some $\rho_1, \ldots, \rho_m, \sigma_1, \ldots, \sigma_n \in [0,1]$ and $a_1, \ldots, a_m, b_1, \ldots, b_n \in Y$. And so, since by Definition B.1

$$f \precsim g \quad \Longleftrightarrow \quad \sum_{i=1}^{n} \rho_i a_i \precsim \sum_{i=1}^{n} \sigma_j b_j,$$

(B.6) and the fact that v is a utility on L imply

$$f \precsim g \quad \Longleftrightarrow \quad \mathcal{E}[u(f)] \leq \mathcal{E}[u(g)]$$

which is the desired result.

I will prove (B.6) by induction on m. First I will show that (B.6) holds for $m = 3$. (Of course, if (B.6) holds for $m = 3$, then it follows that (B.6) also holds for $m = 1$ and $m = 2$, since we can allow some or all of the a_i appearing in (B.6) to be identical.) Suppose then that $\sum_{i=1}^{3} \rho_i a_i \in L$. It follows that for $i = 1, 2, 3$,

$$\rho_i a_i + (1 - \rho_i)a \in L$$

and so

$$\sum_{i=1}^{3} (1/3)\rho_i a_i + (2/3)a = \sum_{i=1}^{3} (1/3)[\rho_i a_i + (1 - \rho_i)a] \in L.$$

Therefore

$$\begin{aligned} v\left(\sum_{i=1}^{3} \rho_i a_i\right) &= \sum_{i=1}^{3} v[\rho_i a_i + (1 - \rho_i)a] \\ &= \sum_{i=1}^{3} \rho_i u(a_i). \end{aligned}$$

Thus (B.6) holds for $m = 3$.

Now to complete the induction, suppose (B.6) is true for some m, and let $\sum_{i=1}^{m+1} \rho_i a_i \in \mathbf{L}$. Define $H_1, H_2 \in \mathbf{L}$ by

$$H_1 = \sum_{i=1}^{m-1} \rho_i a_i + (\rho_m + \rho_{m+1})a$$

$$H_2 = \rho_m a_m + \rho_{m+1} a_{m+1} + \left(\sum_{i=1}^{m-1} \rho_i\right) a.$$

Then by assumption,

$$v(H_1) = \sum_{i=1}^{m-1} \rho_i u(a_i)$$

$$v(H_2) = \rho_m u(a_m) + \rho_{m+1} u(a_{m+1}).$$

But

$$\sum_{i=1}^{m+1} (\rho_i/2) a_i + (1/2)a = (1/2)H_1 + (1/2)H_2$$

and so

$$v\left(\sum_{i=1}^{m+1} \rho_i a_i\right) = v(H_1) + v(H_2) = \sum_{i=1}^{m+1} \rho_i u(a_i).$$

Thus (B.6) is true with $m + 1$ in place of m. ∎

Appendix C

Sufficient conditions for Axiom 10

This appendix proves that when Axioms 1–8 are satisfied, then the following conditions are jointly sufficient (but not necessary) for Axiom 10:

(i) There are "best" and "worst" acts; that is, there exist $k, l \in \mathbf{D}$ such that, for all $f \in \mathbf{D}$, $k \precsim f \precsim l$.

(ii) For all $f \in \mathbf{D}$, if $k \prec f$ or $f \prec l$, then there exists $h \in \bigcup \mathbf{L}$ such that $h \precsim f$ or $f \precsim h$ (respectively).

As in Appendix B, we here use the fact that the set of simple acts in $\mathbf{D}$ is $\bigcup \mathbf{L}$.

Suppose conditions (i) and (ii) are satisfied, and let $f, g \in \mathbf{D}$ be such that $f \prec g$. I will show that there exists $h \in \bigcup \mathbf{L}$ such that $f \precsim h \precsim g$. First note that, by (ii), there exist $h_1, h_2 \in \bigcup \mathbf{L}$ such that $h_1 \prec g$ and $f \prec h_2$. If either $f \precsim h_1$ or $h_2 \precsim g$, there would be nothing more to prove; so suppose that these conditions do not hold; that is, suppose $h_1 \prec f \prec g \prec h_2$. Let

$$\alpha = \sup \left\{ \mathcal{E}[u(h)] : h \in \bigcup \mathbf{L}, \ h \prec f \right\}$$
$$\beta = \inf \left\{ \mathcal{E}[u(h)] : h \in \bigcup \mathbf{L}, \ g \prec h \right\}.$$

Obviously $\alpha \leq \beta$. Now since $\mathbf{L}$ is a convex set (Theorem B.4), there exists $h \in \bigcup \mathbf{L}$ such that

$$\mathcal{E}[u(h)] = (\alpha + \beta)/2.$$

But then if $\alpha < \beta$, we would have

$$\alpha < \mathcal{E}[u(h)] < \beta,$$

265

and hence $f \prec h \prec g$, and the result would be established. So suppose instead that $\alpha = \beta$. Then we have

$$\mathcal{E}[u(h)] = \alpha = \beta.$$

Now if $h \prec f$, then $h \prec h_2$, and so by Axiom 8 there exists $h' \in \bigcup \mathbf{L}$ such that $h \prec h' \prec f$. But then $\mathcal{E}[u(h')] > \mathcal{E}[u(h)] = \alpha$, contradicting the choice of α. Hence $f \precsim h$. A similar argument shows that $h \precsim g$, and so we have the desired result that $f \precsim h \precsim g$.

Appendix D

Proof of Theorem 8.2

This appendix gives a proof of Theorem 8.2, which establishes an expected utility representation for preferences over cognitive acts that are not necessarily simple. As in Appendix B, 'p' denotes the unique probability function whose existence is vouched for by Theorem B.1. Also, it is assumed throughout this appendix that Axioms 1–11 are satisfied.

D.1 COUNTABLE ADDITIVITY OF PROBABILITY

In this section, it will be shown that p is countably additive – that is, that for all disjoint sequences $\{A_i\}$ in $\mathbf{X}$, $p(\bigcup_i A_i) = \sum_i p(A_i)$. The proof of that result will utilize the following theorem.

Theorem D.1. *If p is not countably additive, then there exists $A \in \mathbf{X}$ and a sequence $\{C_i\}$ in $\mathbf{X} \setminus \mathbf{N}$ such that $p(A) > 1/2$, $\{C_i\}$ is a partition of X, and $p(A|C_i) < 1/2$ for all $i \in I^+$.*

Proof.[1] Let us say that a probability p is *dilute* if for all $\varepsilon > 0$ there exists a measurable partition $\{D_i\}$ of $\mathbf{X}$ such that $\sum_i p(D_i) < \varepsilon$. Then it follows from Theorems 2.1 and 2.2 of Schervish, Seidenfeld, and Kadane (1984) that any probability is a convex combination of a countably additive probability and a dilute probability. Thus for any probability p, we can write

$$p = \alpha p_c + \beta p_d$$

where p_c is a countably additive probability, p_d is a dilute probability, $\alpha \in [0,1]$, and $\beta = 1 - \alpha$.

Now suppose that p is not countably additive, so that $\beta > 0$ in the above decomposition. By the continuity of p (Theorem B.1,

---

[1]This proof is due to Teddy Seidenfeld (personal communication, April 5, 1984).

part iv), there is a partition $\{E_1, E_2, E_3\}$ of X such that $p(E_1) = p(E_2) = 0.5 - 0.1\beta$. Also, by renaming E_1 and E_2 if necessary, we can suppose that $p_d(E_1) \geq p_d(E_2)$. Letting $A = E_1 \cup E_3$, we then have that $p(A) = 0.5 + 0.1\beta$ and $p_d(A) \geq 0.5$. By the definition of a dilute probability, there exists a partition $\{A_i\}$ of A such that $\sum_i p_d(A_i) < 0.01$. Then

$$
\begin{aligned}
\sum_i p(A_i) &= \alpha p_c(A) + \beta \sum_i p_d(A_i) \\
&= p(A) - \beta p_d(A) + \beta \sum_i p_d(A_i) \\
&< 0.5 - 0.39\beta \\
&< 0.5 - 0.1\beta = p(\bar{A}).
\end{aligned}
$$

It follows from this inequality that there is a partition $\{B_i\}$ of $\bar{A}$ such that $p(B_i) > p(A_i)$ for all $i \in I^+$. For by the continuity of p, there is a disjoint sequence $\{B_i\}$ of subsets of $\bar{A}$ such that, for all $i \in I^+$,

$$
p(B_i) = p(A_i) + \frac{p(\bar{A}) - \sum_i p(A_i)}{2^i}.
$$

Finally, for each $i \in I^+$, let $C_i = A_i \cup B_i$. Then we have

$$
p(A|C_i) = \frac{p(A_i)}{p(A_i) + p(B_i)} < 1/2. \quad \blacksquare
$$

The countable additivity of p is an easy corollary of Theorem D.1.

Theorem D.2. *p is countably additive.*

Proof. Suppose p is not countably additive. Then by Theorem D.1 there exists $A \in \mathbf{X}$ such that $p(A) > 1/2$, and there exists a sequence $\{C_i\}$ in $X \setminus \mathbf{N}$ such that $\{C_i\}$ is a partition of X and $p(A|C_i) < 1/2$ for all $i \in I^+$. Let $a, b \in Y$ be such that $a \prec b$, and define $f, g \in \bigcup \mathbf{L}$ by the conditions

$$
f = \left\{ \begin{array}{l} b \text{ on } A \\ a \text{ on } \bar{A} \end{array} \right\}, \qquad g = \left\{ \begin{array}{l} a \text{ on } A \\ b \text{ on } \bar{A} \end{array} \right\}.
$$

Let u be a utility on Y^* (Definition B.3). Then $u(a) < u(b)$. Also for all $i \in I^+$, if

$$g' = \left\{ \begin{array}{l} g \text{ on } C_i \\ f \text{ on } \bar{C}_i \end{array} \right\}$$

then $\mathcal{E}[u(f)] < \mathcal{E}[u(g')]$, and so $f \prec g'$; that is, $f \prec g$ given C_i. But $\mathcal{E}[u(f)] > \mathcal{E}[u(g)]$, and so $f \succ g$, violating Axiom 9. Hence it is false that p is not countably additive. ∎

D.2 THE FUNCTION w

In this section I will define a function w with domain $\mathbf{D}$ and establish some basic properties of this function – properties that will justify thinking of w as a kind of utility function on $\mathbf{D}$. But before turning to that, I will establish several useful results, beginning with the following strengthening of Axiom 10.

Theorem D.3. *If $f, g \in \mathbf{D}$ and $f \prec g$, then there exists $h \in \bigcup \mathbf{L}$ such that $f \prec h \prec g$.*

Proof. Let $a, b \in Y$ be such that $a \prec b$. Then at least one of the following must hold:

(i) $f \prec b$.
(ii) $a \prec g$.

Suppose (i) holds. Then by Axiom 8, there exists a measurable partition $\{A_1, \ldots, A_n\}$ of X such that for all $i = 1, \ldots, n$, if

$$f_i = \left\{ \begin{array}{l} b \text{ on } A_i \\ f \text{ on } \bar{A}_i \end{array} \right\}$$

then $f_i \prec g$. But by Theorem B.2, there exists $j \in \{1, \ldots, n\}$ such that $f \prec b$ given A_j, whence $f \prec f_j \prec g$. Also, repeating the preceding argument with f_j in place of g, we have that there exists $f'_j \in \mathbf{D}$ such that $f \prec f'_j \prec f_j$. Then by Axiom 10 there exists $h \in \bigcup \mathbf{L}$ such that $f'_j \precsim h \precsim f_j$, whence $f \prec h \prec g$.

269

If (ii) holds, then using Axiom 8 and Theorem B.2 much as before, we have that there exist $g_j, g'_j \in \mathbf{D}$ such that $f \prec g_j \prec g'_j \prec g$. Then by Axiom 10 there exists $h \in \bigcup \mathbf{L}$ such that $g_j \precsim h \precsim g'_j$, whence $f \prec h \prec g$ again. ∎

The following theorem is proved by Savage (1954, p. 73). His proof of it is unaffected by the differences between his postulates and my axioms, and so the theorem is stated here without proof.

Theorem D.4. *If $F_1, F_2 \in \mathbf{L}$, $g \in \mathbf{D}$, and $F_1 \prec g \prec F_2$, then there exists a unique $\rho \in [0, 1]$ such that $\rho F_1 + (1 - \rho)F_2 \sim g$.*

Let $a_0 \in Y$ be such that $a_0 \prec a$ for some $a \in Y$. By Theorems B.11 and B.12, there exists a function u_0 such that u_0 is a utility on Y^* (Definition B.3) and $u_0(a_0) = 0$. We keep a_0 and u_0 fixed throughout the remainder of this appendix. Also when $g \in \bigcup \mathbf{L}$ I will for brevity write $u_0(g)$ in place of $\mathcal{E}[u_0(g)]$, thus regarding u_0 as a function on $\bigcup \mathbf{L}$ as well as on Y. No confusion should arise from this convention, since these two uses of the symbol u_0 are distinguished by the different arguments that the function takes in each case. In the following theorem, for example, it is clear that u_0 is being regarded as a function on $\bigcup \mathbf{L}$.

Theorem D.5. *For all $f \in \mathbf{D}$ there exists $g \in \bigcup \mathbf{L}$ such that $g \precsim f$ or $f \precsim g$. If there exist $g, h \in \bigcup \mathbf{L}$ such that $g \precsim f \precsim h$, then*

$$\sup \left\{ u_0(g) : \ g \in \bigcup \mathbf{L}, \ g \precsim f \right\}$$
$$= \inf \left\{ u_0(h) : \ h \in \bigcup \mathbf{L}, \ f \precsim h \right\}. \qquad \text{(D.1)}$$

Proof. The first part of the theorem follows immediately from the fact that $\precsim$ is connected (Axiom 1) and $\bigcup \mathbf{L} \neq \emptyset$ (Axiom 2).

As for the second part, we have by Theorem D.4 that there always exists $k \in \bigcup \mathbf{L}$ such that $f \sim k$. Thus

$$\sup \left\{ u_0(g) : \ g \in \bigcup \mathbf{L}, \ g \precsim f \right\}$$
$$\geq \inf \left\{ u_0(h) : \ h \in \bigcup \mathbf{L}, \ f \precsim h \right\},$$

270

since k is in both sets. But the reverse inequality follows from the fact that u_0 is order preserving on $\bigcup \mathbf{L}$, and so (D.1) holds. ∎

In view of Theorem D.5, we can define a function w on $\mathbf{D}$ as follows.

Definition D.1. *For all $f \in \mathbf{D}$, if there exists $g \in \bigcup \mathbf{L}$ such that $g \precsim f$, then*

$$w(f) = \sup \left\{ u_0(g) : g \in \bigcup \mathbf{L}, \ g \precsim f \right\}$$

and if there exists $h \in \bigcup \mathbf{L}$ such that $f \precsim h$, then

$$w(f) = \inf \left\{ u_0(h) : h \in \bigcup \mathbf{L}, \ f \precsim h \right\}.$$

It is clear from Definition D.1 that w is an extended real-valued function; that is, $w(f) \in [-\infty, \infty]$ for all $f \in \mathbf{D}$. Also w agrees with u_0 on $\bigcup \mathbf{L}$; that is, $w(g) = u_0(g)$ for all $g \in \bigcup \mathbf{L}$. So since u_0 is finite and order preserving on $\bigcup \mathbf{L}$, the same holds for w. The following two theorems generalize this result, showing that w is in fact order preserving and finite on $\mathbf{D}$.

Theorem D.6. *For all $f, g \in \mathbf{D}$, $f \precsim g$ iff $w(f) \le w(g)$.*

Proof. If $f \sim g$, it follows immediately from Definition D.1 that $w(f) = w(g)$. Suppose then that $f \prec g$. Applying Theorem D.3 twice, we have that there exist $h_1, h_2 \in \bigcup \mathbf{L}$ such that $f \prec h_1 \prec h_2 \prec h$. But then Definition D.1 gives that

$$w(f) \le w(h_1) < w(h_2) \le w(g).$$

Thus we have shown that if $f \precsim g$, then $w(f) \le g$.

Reversing the role of f and g in the preceding argument, we have that if $g \prec f$ then $w(g) < w(f)$. So by contraposition, if $w(f) \le w(g)$, then $f \precsim g$. ∎

Before turning to the proof of the finiteness of w, I introduce the following notation, which will be used both in that proof and subsequently.

271

Definition D.2. *For all $f \in \mathbf{D}$ and $A \in \mathbf{X}$, f_A is the unique element of $\mathbf{D}$ such that*

$$f_A = \left\{ \begin{array}{c} f \text{ on } A \\ a_0 \text{ on } \bar{A} \end{array} \right\}.$$

The finiteness of w may now be proved as follows.[2]

Theorem D.7. *w is finite.*

Proof. Suppose there is an f in $\mathbf{D}$ such that $w(f) = \infty$. Then Definition D.1 gives that

$$\sup\{u_0(g) : g \in \bigcup \mathbf{L}\} = \infty. \tag{D.2}$$

Let $A \in \mathbf{X}$ be such that $A, \bar{A} \notin \mathbf{N}$. It is easy to verify that for all $g \in \bigcup \mathbf{L}$,

$$u_0(g) = u_0(g_A) + u_0(g_{\bar{A}}).$$

So we can infer from (D.2) that either

$$\sup\{u_0(g_A) : g \in \bigcup \mathbf{L}\} = \infty \tag{D.3}$$

or

$$\sup\{u_0(g_{\bar{A}}) : g \in \bigcup \mathbf{L}\} = \infty$$

(or both). Suppose without loss of generality that (D.3) holds. Then since u_0 is finite on $\bigcup \mathbf{L}$, we have that for all $g \in \bigcup \mathbf{L}$ there exists $h \in \bigcup \mathbf{L}$ such that $u_0(h_A) > u_0(g_A)$, and hence $h_A \succ g_A$. So if $f_A \precsim g_A$ for some $g \in \bigcup \mathbf{L}$, then $f_A \prec h_A$ for some $h \in \bigcup \mathbf{L}$, and defining f' by the condition

$$f' = \left\{ \begin{array}{c} h_A \text{ on } A \\ f \text{ on } \bar{A} \end{array} \right\}$$

gives that $f' \succ f$ (by Theorem B.2). But this violates Theorem D.6, since $w(f') \leq \infty = w(f)$. Hence $f_A \succ g_A$ for all $g \in \bigcup \mathbf{L}$, and consequently $w(f_A) = \infty$.

[2] A stronger result, namely that w is bounded, has been proved by Teddy Seidenfeld (personal communication, June 8, 1983).

Now by our choice of a_0, there exists $a \in Y$ such that $a_0 \prec a$. So if we let

$$f'' = \left\{ \begin{array}{l} f \text{ on } A \\ a \text{ on } \bar{A} \end{array} \right\}$$

we have (by Theorem B.2) that $f'' \succ f_A$. But this violates Theorem D.6, since $w(f'') \leq \infty = w(f_A)$. Hence our original assumption, that $w(f) = \infty$, is false. A similar argument shows that $w(f) \neq -\infty$. ∎

D.3 THE SIGNED MEASURES w_f

I will now define, for each act f in $\mathbf{D}$, a real-valued function w_f on $\mathbf{X}$. One can think of $w_f(A)$, for any $A \in \mathbf{X}$, as a measure of the expected utility of f when the event A is known to occur. The formal definition is as follows.

Definition D.3. *For all $f \in \mathbf{D}$, w_f is the function defined on $\mathbf{X}$ by the condition that for all $A \in \mathbf{X}$,*

$$w_f(A) = w(f_A).$$

This section will be devoted to establishing some properties of the functions w_f just defined. One property that the reader can easily verify – and which will be appealed to in subsequent proofs – is that w_f is finitely additive when $f \in \bigcup \mathbf{L}$.

The following theorem shows that we can replace f, g, and h in Definition D.1 by f_A, g_A, and h_A, for any $A \in \mathbf{X}$.

Theorem D.8. *Let $f \in \mathbf{D}$ and $A \in \mathbf{X}$. If there exists $g \in \bigcup \mathbf{L}$ such that $g_A \precsim f_A$, then*

$$w(f_A) = \sup \left\{ u_0(g_A) : g \in \bigcup \mathbf{L}, \ g_A \precsim f_A \right\}$$

and if there exists $g \in \bigcup \mathbf{L}$ such that $f_A \precsim h_A$, then

$$w(f_A) = \inf \left\{ u_0(h_A) : h \in \bigcup \mathbf{L}, \ f_A \precsim h_A \right\}.$$

273

Proof. If $p(A) = 1$, then for all $g \in \bigcup \mathbf{L}$, $g_A \precsim f_A$ iff $g \precsim f$, and $u_0(g_A) = u_0(g)$. The desired result then follows directly from Definition D.1.

Suppose there exist $g, h \in \bigcup \mathbf{L}$ such that $g_A \precsim f_A \precsim h_A$. Then if $G, H \in \mathbf{L}$ are such that $g_A \in G$ and $h_A \in H$, we have by Theorem D.4 that for some $\rho \in [0, 1]$,

$$\rho G + (1 - \rho)H \sim f_A.$$

Let $B \in \mathbf{X}$ be such that for all $y \in Y$,

$$p\left[B \cap g_A^{-1}(y) \cap h_A^{-1}(y)\right] = \rho\, p\left[g_A^{-1}(y) \cap h_A^{-1}(y)\right].$$

Define $k \in \bigcup \mathbf{L}$ by the condition

$$k = \left\{ \begin{array}{c} g \text{ on } B \\ h \text{ on } \bar{B} \end{array} \right\}.$$

Then $k_A \in \rho G + (1 - \rho)H$, and so $k_A \sim f_A$. The desired result now follows from the fact that u_0 is order preserving on $\bigcup \mathbf{L}$ and

$$u_0(k_A) = w(k_A) = w(f_A).$$

Two cases remain to be considered:

(i) $p(A) < 1$ and for all $g \in \bigcup \mathbf{L}$, $g_A \prec f_A$.
(ii) $p(A) < 1$ and for all $g \in \bigcup \mathbf{L}$, $f_A \prec g_A$.

Suppose (i) holds, and for all $E \in \mathbf{X}$ let

$$\hat{w}_E = \sup\{w(g_E) : g \in \bigcup \mathbf{L}\}.$$

By Theorem D.7, $w(f_A)$ is finite, and so since $0 \leq \hat{w}_A \leq w(f_A)$, $\hat{w}_A$ is also finite. Let $\varepsilon = w(f_A) - \hat{w}_A$, and suppose $\varepsilon > 0$. Choose $k \in \mathbf{D}$ such that

$$w(k_{\bar{A}}) \geq \max\left\{\hat{w}_{\bar{A}} - \frac{\varepsilon}{2}, 0\right\}$$

274

and define $f' \in \mathbf{D}$ by the condition

$$f' = \left\{ \begin{array}{l} f \text{ on } A \\ k \text{ on } \bar{A} \end{array} \right\}.$$

By Definition D.1, there exists $g \in \bigcup \mathbf{L}$ such that

$$w(f_A) - \varepsilon < w(g) \leq w(f_A).$$

Since $f_A \succ a$, the weak inequality here can be replaced with a strict one; that is, there exists $g \in \bigcup \mathbf{L}$ such that

$$w(f_A) - \varepsilon < w(g) < w(f_A).$$

For any such g, there exists by Axiom 8 an $f'' \in \mathbf{D}$ such that $g \prec f''_A \prec f_A$. But this means

$$w(f_A) - \varepsilon < w(f''_A) < w(f_A).$$

Since $w(f_A) - \varepsilon = \hat{w}_A$, we then have that for all $g \in \bigcup \mathbf{L}$, $f'' \succ g$ given A.

Now if there exists $h \in \mathbf{D}$ such that $w(h_{\bar{A}}) \geq \hat{w}_{\bar{A}}$, then setting

$$k = \left\{ \begin{array}{l} f \text{ on } A \\ h \text{ on } \bar{A} \end{array} \right\}, \qquad k' = \left\{ \begin{array}{l} f'' \text{ on } A \\ h \text{ on } \bar{A} \end{array} \right\}$$

we have that for all $g \in \bigcup \mathbf{L}$, $k \succ k' \succ g$ given A, and $k \sim k' \succsim g$ given $\bar{A}$, whence $k \succ k' \succ g$. But this violates Axiom 10. Hence for all $h \in \mathbf{D}$, $w(h_{\bar{A}}) < \hat{w}_{\bar{A}}$. In particular, $w(f'_{\bar{A}}) < \hat{w}_{\bar{A}}$. Consequently, there exists $h \in \mathbf{D}$ such that $w(f'_{\bar{A}}) < w(h_{\bar{A}})$. Then setting

$$k = \left\{ \begin{array}{l} f \text{ on } A \\ h \text{ on } \bar{A} \end{array} \right\}$$

gives $k \succ f'$. By Axiom 10 we then have that there exists $g \in \bigcup \mathbf{L}$ such that $f' \precsim g \precsim k$. Also $a \prec f'$, and so by an application of Theorem D.4 we infer that there exists $g \in \bigcup \mathbf{L}$ such that $g \sim f'$.

By the assumption (i), $g_A \prec f_A$. Also since $f' = f_A$ on A and $f' \succsim f_A$ given $\bar{A}$, we have that $f' \succsim f'_A$, and so the fact that $g \sim f'$ implies $g \succsim f'_A = f_A$. In short, $g_A \prec f_A \precsim g$. It follows that there exists $B \in \mathbf{X}$ such that $B \supset A$ and $g_B \sim f_A$. Also since w_g is finitely additive,

$$w(g_{B \setminus A}) = w(g_B) - w(g_A) = w(f_A) - w(g_A) \geq \varepsilon. \qquad \text{(D.4)}$$

Now define $g' \in \mathbf{D}$ by the condition

$$g' = \left\{ \begin{array}{c} f_A \text{ on } B \\ g \text{ on } \bar{B} \end{array} \right\}.$$

Then $g' \sim g$ given B and $g' \sim g$ given $\bar{B}$, whence $g' \sim g \sim f'$. But the fact that $f_A = f'_A$ implies that $g' = f'$ on A, and so we have that $g' \sim f'$ given $\bar{A}$, whence $g'_{\bar{A}} = f'_{\bar{A}}$. Also since $g' = a$ on $B \setminus A$, we have $g'_{\bar{A}} = g'_{\bar{B}} = g_{\bar{B}}$. Thus $g_{\bar{B}} \sim f'_A$, whence

$$w(g_{\bar{B}}) = w(f'_A) \geq \max\{\hat{w}_{\bar{A}} - \frac{\varepsilon}{2}, 0\} > \hat{w}_{\bar{A}} - \varepsilon. \qquad \text{(D.5)}$$

Using (D.4) and (D.5) together with the additivity of w_g, we obtain

$$w(g_{\bar{A}}) = w(g_{B \setminus A}) + w(g_{\bar{B}}) > \hat{w}_{\bar{A}}.$$

But this contradicts the definition of $\hat{w}_{\bar{A}}$, and so the assumption that $\varepsilon > 0$ is false, whence $\varepsilon = 0$. That is, $w(f_A) = \hat{w}_A$, and so the theorem holds in this case.

The proof that the theorem holds in case (ii) is similar to the proof just given with regard to (i). ∎

Theorem D.9. *For all $f \in \mathbf{D}$, w_f is finitely additive.*

Proof. Let $f \in \mathbf{D}$, and let $\{A_1, A_2\}$ be a measurable partition of $A \in \mathbf{X}$. Let

$$\varepsilon = \sum_{i=1}^{2} w_f(A_i) - w_f(A).$$

We want to show that $\varepsilon = 0$.

The following four cases are mutually exhaustive:

(i) There exists $g \in \bigcup \mathbf{L}$ such that $g_{A_1} \precsim f_{A_1}$ and $g_{A_2} \precsim f_{A_2}$.

(ii) For all $g \in \bigcup \mathbf{L}$, $g_{A_1} \succsim f_{A_1}$ and $g_{A_2} \succsim f_{A_2}$.

(iii) There exists $g \in \bigcup \mathbf{L}$ such that $g_{A_1} \prec f_{A_1}$ and, for all $h \in \bigcup \mathbf{L}$, $h_{A_2} \succ f_{A_2}$.

(iv) There exists $g \in \bigcup \mathbf{L}$ such that $g_{A_2} \prec f_{A_2}$ and, for all $h \in \bigcup \mathbf{L}$, $h_{A_1} \succ f_{A_1}$.

I now suppose that $\varepsilon > 0$, and proceed to show that on this supposition each of (i)–(iv) is impossible (from which we will be able to conclude that $\varepsilon \leq 0$).

If (i) holds, then by Theorem D.8 there exists $g \in \bigcup \mathbf{L}$ such that for $i = 1, 2$:

$$w_f(A_i) - \frac{\varepsilon}{2} < w_g(A_i) \leq w_f(A_i).$$

Then since w_g is additive, $w_g(A) > \sum_{i=1}^{2} w_f(A_i) - \varepsilon = w_f(A)$. Thus $g \succ f$ given A, although $g \precsim f$ given A_1 and $g \precsim f$ given A_2. But this contradicts Theorem B.2, and so (i) is impossible.

If (ii) holds, then

$$\inf \left\{ w_g(A) : \ g \in \bigcup \mathbf{L} \right\} \geq \sum_{i=1}^{2} w_f(A_i) > w_f(A),$$

which contradicts Theorem D.8. So (ii) is impossible.

If (iii) holds, then by Theorem D.8 there exists $g \in \bigcup \mathbf{L}$ such that

$$w_f(A_1) - \varepsilon < w_g(A_1) < w_f(A_1)$$

and (setting $\delta = w_f(A_1) - w_g(A_1)$)

$$w_f(A_2) < w_g(A_2) < w_f(A_2) + \delta.$$

Then since

$$w_g(A) = \sum_{i=1}^{2} w_g(A_i) > \sum_{i=1}^{2} w_f(A_i) - \varepsilon = w_f(A)$$

277

we have that $g_A \succ f_A$. So by Axiom 8, there is a measurable partition $\{B_1, \ldots, B_n\}$ of A_2 such that for $i = 1, \ldots, n$, if

$$f_i' = \left\{ \begin{array}{l} a_0 \text{ on } B_i \\ f \text{ on } \bar{B}_i \end{array} \right\}$$

then $g_A \succ f_A'$. Now $f_{A_2} \prec a_0$ by assumption, and so there exists $k \in \{1, \ldots, n\}$ such that $f_k' \succ f$. Then letting $f' = f_k'$, we have $g_A \succ f_A' \succ f_A$. Also $f_{A_2}' \succ f_{A_2}$, so using Theorem D.8 again we can find $g' \in \bigcup \mathbf{L}$ such that $g' = g$ on A_1, $g' \precsim g$ given A_2, and $g' \prec f'$ given A_2. Then $w_g(A_2) \geq w_{g'}(A_2)$ and

$$\begin{aligned} w_g(A_1) + w_g(A_2) - w_{g'}(A_2) \; &< \; w_g(A_1) + w_g(A_2) - w_f(A_2) \\ &< \; w_g(A_1) + \delta \\ &< \; w_f(A_1). \end{aligned}$$

So, by a further application of Theorem D.8, there exists $g'' \in \bigcup \mathbf{L}$ such that $g'' = g'$ on A_2 and

$$w_{g''}(A_1) = w_g(A_1) + w_g(A_2) - w_{g'}(A_2).$$

Then $w_{g''}(A_1) = w_g(A)$, and so $g_A'' \sim g_A$, whence $g_A'' \succ f_A'$. But $g_{A_1}'' \prec f_{A_1} = f_{A_1}'$, and $g_{A_2}'' = g_{A_2}' \prec f_{A_2}'$, so that Axiom 8 is violated. Consequently, (iii) is impossible.

To see that (iv) is impossible, one need only interchange A_1 and A_2 in the proof that (iii) is impossible.

This completes the proof that $\varepsilon \leq 0$. The proof that $\varepsilon \geq 0$ is just the dual of the above, and so $\varepsilon = 0$; that is, $w_f(A) = \sum_{i=1}^{2} w_f(A_i)$. It then follows by mathematical induction that w_f is finitely additive. ∎

A real-valued function φ on $\mathbf{X}$ is said to be *absolutely continuous* if for all $\varepsilon > 0$ there exists $\delta > 0$ such that $|\varphi(A)| < \varepsilon$ for all $A \in \mathbf{X}$ for which $p(A) < \delta$. I will now show that the functions w_f have this property.

Theorem D.10. *For all $f \in \mathbf{D}$, w_f is absolutely continuous.*

Proof. If w_f is not absolutely continuous, then there exists $\varepsilon > 0$ such that at least one of the following holds:

278

(i) For all $\delta > 0$ there exists $A \in \mathbf{X}$ such that $p(A) < \delta$ and $w_f(A) > \varepsilon$.

(ii) For all $\delta > 0$ there exists $A \in \mathbf{X}$ such that $p(A) < \delta$ and $w_f(A) < -\varepsilon$.

Suppose (i) holds for some $\varepsilon > 0$. Then there exist sequences $\{A_i\}$ in $\mathbf{X}$, $\{\rho_i\}$ in $(0, \infty)$, and $\{g_i\}$ in $\bigcup \mathbf{L}$ such that for all $i \in I^+$:

(a) $w_f(A_i) > \varepsilon$.

(b) $w_{g_i} > \varepsilon$.

(c) $\rho_i = \varepsilon / \max \left\{ u_0(y) : y \in Y, \ g_i^{-1}(y) \notin \mathbf{N} \right\}$.

(d) $p(A_{i+1}) < \rho_i/4$.

(The existence of a $g_i \in \bigcup \mathbf{L}$ satisfying (b) is guaranteed by Theorem D.8.) Then for all $i \in I^+$ we have by (b) that

$$\varepsilon < p(A_i) \max \left\{ u_0(y) : y \in Y, \ g_i^{-1} \notin \mathbf{N} \right\}.$$

Hence by (c) and (d),

$$4p(A_{i+1}) < \rho_i < p(A_i). \tag{D.6}$$

Now define a sequence $\{B_i\}$ in $\mathbf{X}$ by the condition that for all $i \in I^+$,

$$B_i = A_i - \bigcup_{j=i+1}^{\infty} A_j.$$

Note that in view of (D.6),

$$p\left(\bigcup_{j=i+1}^{\infty} A_j \right) \le \sum_{j=i+1}^{\infty} p(A_j) < 2p(A_{i+1}) < \frac{\rho_i}{2}. \tag{D.7}$$

We then have

$$
\begin{aligned}
w_{g_i}(B_i) &= w_{g_i}(A_i) - w_{g_i}\left(\bigcup_{j=i+1}^{\infty} A_j \right) \\
&\ge w_{g_i}(A_i) - p\left(\bigcup_{j=i+1}^{\infty} A_j \right) \times \\
&\quad \max \left\{ u_0(y) : y \in Y, \ g_i^{-1}(y) \notin \mathbf{N} \right\} \\
&> \frac{\varepsilon}{2}, \quad \text{by (b), (c), and (D.7).}
\end{aligned}
$$

279

Since the B_i are pairwise disjoint, we can define a sequence $\{h_i\}$ in $\bigcup \mathbf{L}$ by the condition

$$h_i = \left\{ \begin{array}{l} g_i \text{ on } B_j, j = 1, \ldots, i \\ a_0 \text{ elsewhere} \end{array} \right\}.$$

Then for all $i \in I^+$,

$$w(h_i) = \sum_{j=1}^{i} w_{g_j}(B_i) > \sum_{j=1}^{i} \frac{\varepsilon}{2} = \frac{i\varepsilon}{2}.$$

Thus $w(h_i) \to \infty$ as $i \to \infty$. Now define $h \in \mathbf{D}$ by the condition

$$h = \left\{ \begin{array}{l} g_j \text{ on } B_j, \ j \in I^+ \\ a_0 \text{ elsewhere} \end{array} \right\}.$$

Then for all $j \in I^+$, $w_h(B_j) = w_{g_j}(B_j) > 0$, and so by Theorem D.6 $h \succ h_i$ given B_j, for all $j > i$. It follows by Axiom 9 that $h \succsim h_i$ given $\bigcup_{j=i+1}^{\infty} B_j$, and so since $h = h_i$ on $\bigcup_{j=1}^{i} B_j$, we have by Theorem B.2 that $h \succsim h_i$. Thus $w(h) \geq w(h_i)$ for all i, whence $w(h) = \infty$. Since this result contradicts Theorem D.7, we infer that (i) cannot hold for any $\varepsilon > 0$.

A slight modification of the preceding argument shows similarly that (ii) also cannot hold for any $\varepsilon > 0$, whence w_f is absolutely continuous. ∎

Having now established that the functions w_f are both finitely additive and absolutely continuous, it follows easily that these functions are in fact countably additive.

Theorem D.11. *For all $f \in \mathbf{D}$, w_f is countably additive.*

Proof. Let $\{A_i\}$ be a disjoint sequence in $\mathbf{X}$, and let $A = \bigcup_{i=1}^{\infty} A_i$. Let $\varepsilon > 0$. Then by Theorem D.10, there exists $\delta > 0$ such that for all $B \in \mathbf{X}$, if $p(B) < \delta$ then $|w_f(B)| < \varepsilon$. And by Theorem D.2, there exists $n \in I^+$ such that $p(\bigcup_{i=n+1}^{\infty} A_i) < \delta$.

Then for all $k \geq n$ we have (using Theorem D.10 and the fact that $p(\bigcup_{i=k+1}^{\infty} A_i) < \delta$):

$$\left| w_f(A) - \sum_{i=1}^{k} w_f(A_i) \right| = \left| w_f\left(\bigcup_{i=k+1}^{\infty} A_i \right) \right| < \varepsilon.$$

Hence

$$\sum_{i=1}^{\infty} w_f(A_i) = \lim_{k} \sum_{i=1}^{k} (A_i) = w_f(A). \quad \blacksquare$$

An extended real-valued function φ on $\mathbf{X}$ is said to be a *signed measure* if φ is countably additive and $\varphi(\emptyset) = 0$. Since it follows from Theorem D.10 that $w_f(\emptyset) = 0$ for all $f \in \mathbf{D}$, Theorem D.11 establishes that the functions w_f are signed measures.

D.4   UTILITY ON Y

I will now define the notion of a utility on Y. Here I use the usual notation for composition of functions, writing $\varphi \circ \psi$ for the composition of functions φ and ψ, that is, the function $\varphi[\psi(\cdot)]$.

Definition D.4. *A real-valued measurable function u with domain Y is a utility on Y iff for all $f, g \in \mathbf{D}$,*
* $u \circ f$ *is measurable, and*
* $f \precsim g$ *iff $\mathcal{E}(u \circ f) \leq \mathcal{E}(u \circ g)$.*

Essentially Definition D.4 says that a utility on Y is a function whose expected value is order preserving on $\mathbf{D}$. The concept of a utility on Y is thus stronger than that of a utility on Y^* (Definition B.3), since the expected value of a utility on Y^* need only be order preserving on $\bigcup \mathbf{L}$.

In this section I will establish the existence of a utility on Y, and prove a uniqueness result for this utility.

Theorem D.12. *There exists a utility on Y.*

Proof. It has been established in Section D.3 that, for all $f \in \mathbf{D}$, the function w_f is an absolutely continuous finite signed

281

measure. Consequently we have by the Radon–Nikodym Theorem (Halmos 1950, pp. 128f.) that for all $f \in \mathbf{D}$ there exists a real-valued measurable function φ on $\mathbf{X}$ such that for all $A \in \mathbf{X}$,

$$w_f(A) = \int_A \varphi\, dp. \tag{D.8}$$

A function φ satisfying (D.8) is called a *Radon–Nikodym derivative* of w_f.

I will now show that for any $h \in E \cup E'$ there exists a Radon–Nikodym derivative φ of w_h such that

(i) For all $y \in Y$, φ is constant on $h^{-1}(y)$.
(ii) For all $y \in Y^*$, $\varphi = u_0(y)$ on $h^{-1}(y)$.

To this end, let φ' be a Radon–Nikodym derivative of w_h. Let $y \in Y$, $A \in \mathbf{X}$, and $A \subset h^{-1}(y)$. Then

$$u_0(h_A) = u_0(y)p(A). \tag{D.9}$$

Also, since $p(A) > 0$ only if $y \in Y^*$, there exists $g \in \bigcup \mathbf{L}$ such that $g = h_A$ a.e.. But then

$$
\begin{aligned}
u_0(h_A) &= u_0(g), &&\text{since } g = h_A \text{ a.e.} \\
&= w(g), &&\text{since } u_0 = w \text{ on } \bigcup \mathbf{L} \\
&= w(h_A), &&\text{since } g \sim h_A.
\end{aligned}
$$

Hence (D.9) yields

$$w_h(A) = u_0(y)p(A). \tag{D.10}$$

From (D.10), and the fact that φ' is a Radon–Nikodym derivative of w_h, it follows that $\varphi' = u_0(y)$ a.e. on $h^{-1}(y)$. Also we have from clause (ii) of Axiom 11 that there are at most countably many $y \in Y$ such that φ' is not constant on $h^{-1}(y)$. So if

$$B = \{x \in X : \varphi'(x) \neq u_0[h(x)],$$
$$\varphi' \text{ not constant on } h^{-1}[h(x)]\},$$

then $B \in \mathbf{N}$. Then defining φ on X by the condition

$$\varphi(x) = \left\{ \begin{array}{l} u_0[h(x)] \text{ on } B \\ \varphi'(x) \text{ on } \bar{B} \end{array} \right\}$$

we have that φ is measurable and $\varphi = \varphi'$ a.e.. Thus φ is a Radon–Nikodym derivative of w_h that satisfies (i). As for (ii), observe that it follows from clauses (i), (iii), and (iv) of Axiom 11 that, if $y \in Y^*$, then either $h^{-1}(y) = \emptyset$ or $p[h^{-1}(y)] > 0$; in the former case (ii) is trivially satisfied, and in the latter case (ii) follows from the fact that $\varphi' = u_0(y)$ a.e. on $h^{-1}(y)$.

Now for each $h \in \mathbf{E} \cup \mathbf{E}'$ we can define a function v_h on $h(X)$ by the condition that for all $x \in X$,

$$v_h[h(x)] = \varphi(x),$$

where φ is a Radon–Nikodym derivative of w_h satisfying conditions (i) and (ii) above. For each of the functions v_h we define another function u_h by the following two conditions:

(a) If $\mathbf{E}'$ is empty, then $u_h = v_h$.
(b) If $h' \in \mathbf{E}'$, then $u_{h'} = v_{h'}$, and for all $h \in \mathbf{E}$,

$$u_h = \left\{ \begin{array}{l} v_h \text{ on } h(X) \setminus h'(X) \\ v_{h'} \text{ on } h(X) \cap h'(X) \end{array} \right\}.$$

In view of clause (i) of Axiom 11, we may define a function u on Y by the condition

$$u = \bigcup \{u_h : h \in \mathbf{E}\}.$$

I will show that u is a utility on Y.

The first step is to show that, for all $f \in \mathbf{D}$, $w \circ f$ is measurable. If $f \in \mathbf{E}'$ there is nothing to prove, since in that case $u \circ f = v_f \circ f$, and the definition of $v_f \circ f$ entails that it is measurable. If $f \in \mathbf{E}$ and $\mathbf{E}'$ is empty, then again there is nothing

283

to prove, since in this case also $u \circ f = v_f \circ f$. If $f \in \mathbf{E}$ and $f' \in \mathbf{E}'$, then

$$u \circ f = \left\{ \begin{array}{l} v_f \circ f \text{ on } f^{-1}[f(X) - f'(X)] \\ v_{f'} \circ f \text{ on } f^{-1}[f(X) \cap f'(X)] \end{array} \right\}.$$

So if $f(X) \cap f'(X) = \emptyset$, then $u \circ f = u_f \circ f$, which is measurable. On the other hand, if $f(X) \cap f'(X) \neq \emptyset$, then by clause (iv) of Axiom 11 there exists $y \in Y$ such that $f(X) \cap f'(X) = \{y\}$; since $\mathbf{Y}$ contains all the singleton sets of Y (Section 8.2.1), it follows that $f(X) \cap f'(X) \in \mathbf{Y}$, and hence $f^{-1}[f(X) \cap f'(X)] \in \mathbf{X}$. Also clause (iv) of Axiom 11 entails that $f = f'$ on $f^{-1}[f(X) \cap f'(X)]$, so that $v_{f'} \circ f$ is identical to the measurable function $v_{f'} \circ f'$ on the measurable set $f^{-1}[f(X) \cap f'(X)]$. Hence $u \circ f$ is measurable if $f \in \mathbf{E}$. Finally, if $f \in \mathbf{D}$, then by clause (iii) of Axiom 11 there exists a sequence $\{f_i\}$ in $\mathbf{E} \cup \mathbf{E}'$ and a measurable partition $\{A_i\}$ of X such that for all $i \in I^+$, $f = f_i$ on A_i. We have shown that each $u \circ f_i$ is measurable, and from this it follows that $u \circ f$ is measurable.

It remains to show that the expected value of u is order preserving on $\mathbf{D}$. Since w is order preserving on $\mathbf{D}$ (Theorem D.6), it suffices to show that $\mathcal{E}(u \circ f) = w(f)$, for all $f \in \mathbf{D}$. To this end, note that if $f \in \mathbf{D}$, and if $\{f_i\}$ and $\{A_i\}$ are the associated sequences asserted to exist by clause (iii) of Axiom 11, then

$$\begin{aligned} w(f) &= w_f(X) \\ &= \sum_{i=1}^{\infty} w_f(A_i), \quad \text{by Theorem D.11} \\ &= \sum_{i=1}^{\infty} w_{f_i}(A_i) \\ &= \sum_{i=1}^{\infty} \int_{A_i} (v_{f_i} \circ f_i) \, dp, \quad \text{by definition of } v_{f_i}. \end{aligned}$$

But $v_{f_i} \circ f_i$ differs from $u_{f_i} \circ f_i$, if at all, only on a singleton set. The continuity of p (part iv of Theorem B.1) entails that

284

the probability of any singleton set is 0. Hence $v_{f_i} \circ f_i = u_{f_i} \circ f_i$ a.e., and so

$$
\begin{aligned}
w(f) &= \sum_{i=1}^{\infty} \int_{A_i} (u_{f_i} \circ f_i)\, dp \\
&= \sum_{i=1}^{\infty} \int_{A_i} (u \circ f)\, dp \\
&= \mathcal{E}(u \circ f). \quad \blacksquare
\end{aligned}
$$

Utilities on Y are not in general unique up to a positive affine transformation. However, the following weaker condition holds.

Theorem D.13. *If u is a utility on Y, then a necessary and sufficient condition for a real-valued function u' on Y to be a utility on Y is that there exist real numbers ρ and σ, $\rho > 0$, such that for all $f \in \mathbf{D}$, $u' \circ f$ is measurable and*

$$
u \circ f = (\rho u' + \sigma) \circ f \ a.e..
$$

Proof. If u' satisfies the stated condition, then for all $f \in \mathbf{D}$ we have

$$
\mathcal{E}(u \circ f) = \rho \mathcal{E}(u' \circ f) + \sigma.
$$

Hence for all $f, g \in \mathbf{D}$,

$$
f \precsim g \Leftrightarrow \mathcal{E}(u \circ f) \leq \mathcal{E}(u \circ g) \Leftrightarrow \mathcal{E}(u' \circ f) \leq \mathcal{E}(u' \circ g).
$$

Thus the stated condition is sufficient for u' to be a utility on Y.

To establish that the condition is necessary, let u and u' be two utilities on Y. Then u and u' are in particular utilities on Y^*, and so it follows from Theorem B.11 that there exist real numbers ρ and σ, $\rho > 0$, such that $u = \rho u' + \sigma$ on Y^*. Setting $u'' = \rho u' + \sigma$, we then have that for all $g \in \bigcup \mathbf{L}$,

$$
u \circ g = u'' \circ g. \tag{D.11}
$$

Now let $f \in \mathbf{D}$ and suppose $g \precsim f$ for some $g \in \bigcup \mathbf{L}$, so that

$$
w(f) = \sup \left\{ u_0(g) : \ g \precsim f, \ g \in \bigcup \mathbf{L} \right\}.
$$

285

Let

$$\varepsilon = \mathcal{E}(u \circ f) - \sup \left\{ \mathcal{E}(u \circ g) : \; g \precsim f, \; g \in \bigcup \mathbf{L} \right\}. \qquad \text{(D.12)}$$

I will show that $\varepsilon = 0$. Since $\varepsilon \geq 0$ simply by the fact that u is a utility on Y, assume that $\varepsilon > 0$. Then if $g \in \bigcup \mathbf{L}$ and $g \precsim f$,

$$\mathcal{E}\left[(u \circ f) - (u \circ g)\right] > 0.$$

Hence if $A = \left[(u \circ f) - (u \circ g)\right]^{-1} (0, \infty)$, then $A \in \mathbf{X} \setminus \mathbf{N}$. Also by the absolute continuity of indefinite integrals (Halmos 1950, p. 97), there exists $\delta > 0$ such that for all $B \in \mathbf{X}$, if $p(B) < \delta$ then

$$\left| \int_B \left[(u \circ f) - (u \circ g)\right] dp \right| < \varepsilon.$$

Then choosing $B \subset A$ such that $0 < p(B) < \delta$ gives

$$0 < \int_B \left[(u \circ f) - (u \circ g)\right] dp < \varepsilon. \qquad \text{(D.13)}$$

Now define $f' \in \mathbf{D}$ by the condition

$$f' = \left\{ \begin{array}{c} g \text{ on } B \\ f \text{ on } \bar{B} \end{array} \right\}.$$

Then

$$\mathcal{E}(u \circ f') = \mathcal{E}(u \circ f) - \int_B \left[(u \circ f) - (u \circ g)\right] dp$$

and so by (D.13),

$$\mathcal{E}(u \circ f) - \varepsilon < \mathcal{E}(u \circ f') < \mathcal{E}(u \circ f). \qquad \text{(D.14)}$$

Now fix f' as a particular act satisfying (D.14). Then in view of (D.12) we have, for all $g \in \bigcup \mathbf{L}$, that if $g \precsim f$ then

$$g \precsim f' \prec f.$$

Since this violates Axiom 10, we infer that $\varepsilon = 0$; that is,

$$\mathcal{E}(u \circ f) = \sup \left\{ \mathcal{E}(u \circ g) : \; g \precsim f, \; g \in \bigcup \mathbf{L} \right\}. \qquad \text{(D.15)}$$

Obviously (D.15) also holds with u'' in place of u. Furthermore, we have from (D.11) that

$$\sup \{\mathcal{E}(u \circ g) : g \precsim f, g \in \bigcup \mathbf{L}\}$$
$$= \sup \{\mathcal{E}(u'' \circ g) : g \precsim f, g \in \bigcup \mathbf{L}\}.$$

Consequently,
$$\mathcal{E}(u \circ f) = \mathcal{E}(u'' \circ f). \tag{D.16}$$

This derivation of (D.16) has been made on the assumption that there is a $g \in \bigcup \mathbf{L}$ such that $g \precsim f$. If that assumption does not hold, then

$$w(f) = \inf \left\{ u_0(g) : f \precsim g, g \in \bigcup \mathbf{L} \right\},$$

and a simple modification of the previous argument for (D.16) shows that (D.16) holds in this case too. Hence (D.16) holds for all $f \in \mathbf{D}$.

Now for all $C \in \mathbf{X}$ we have

$$\int_C (u \circ f)\, dp = \mathcal{E}(u \circ f_C) - u(a_0)p(\bar{C})$$
$$= \mathcal{E}(u'' \circ f_C) - u''(a_0)p(\bar{C}), \text{ by (D.11) and (D.16)}$$
$$= \int_C (u'' \circ f)\, dp.$$

Hence $u \circ f = u'' \circ f$ a.e.. ∎

D.5 THE NEED FOR AXIOM 11

I conclude this appendix with an example that shows Theorem D.12 does not hold in the absence of Axiom 11.

Let X be the interval $[-1, 1]$, and let $\mathbf{X}$ be the Borel subsets of X. Let $Y = [0, 1]^2 \cup \{0, 1\}$, and let $\mathbf{Y}$ be the set of all A such that $A \setminus \{0, 1\}$ is a Borel subset of $[0, 1]^2$. Let $\mathbf{D}$ be the set of all functions from X to Y such that for some measurable partition $\{A_1, A_2, A_3, A_4\}$ of X:

(i) $f = 0$ on A_1.
(ii) $f = 1$ on A_2.

(iii) There is a measurable partition $\{B_1, \ldots, B_n\}$ of A_3 such that for all $i = 1, \ldots, n$ there exists $y_i \in [0, 1]$ such that for all $x \in B_i$, $f(x) = (y_i, x + y_i)$.

(iv) There is a measurable partition $\{B_1, \ldots, B_n\}$ of A_4 such that for all $i = 1, \ldots, n$ there exists $y_i \in [0, 1]$ such that for all $x \in B_i$, $f(x) = (y_i - x, y_i)$.

(Conditions (iii) and (iv) imply that $f(A_3)$ and $f(A_4)$ are subsets of $[0, 1]^2$. In fact, $f(A_3)$ is a finite union of sets of the form $\{y\} \times C$, where $y \in [0, 1]$ and $C \subset [0, 1]$; similarly $f(A_4)$ is a finite union of sets of the form $C \times \{y\}$. Note also that for all $(y_1, y_2) \in [0, 1]^2$, $f^{-1}(y_1, y_2)$ contains at most the single point $y_2 - y_1$.)

Obviously A_1 and A_2 are uniquely determined by conditions (i) and (ii). Also A_3 and A_4 are determined to within a finite set by conditions (iii) and (iv). For suppose $\{A_1, A_2, A_3, A_4\}$ and $\{A_1, A_2, A_3', A_4'\}$ are two measurable partitions of X satisfying (i)–(iv). A_3 is a finite union of sets of the form

$$f^{-1}(\{y_1\} \times C_1)$$

and A_4 is a finite union of sets of the form

$$f^{-1}(C_2 \times \{y_2\}).$$

Hence $A_3 \cap A_4'$ is a finite union of sets of the form

$$f^{-1}\{(y_1, y_2)\}.$$

But $f^{-1}\{(y_1, y_2)\}$ is a singleton set, and so $A_3 \cap A_4'$ is finite. Hence $A_3 \setminus A_3'$ is finite. Similar reasoning shows that $A_3' \setminus A_3$, $A_4 \setminus A_4'$, and $A_4' \setminus A_4$ are also finite, so that A_3 and A_4 are determined to within a finite set, as claimed.

Since the Lebesgue measure of any finite set is zero, the Lebesgue measure of the sets A_1, A_2, A_3, and A_4 is uniquely determined. Hence we may define a function Q on $\mathbf{D}$ by the condition that, for all $f \in \mathbf{D}$, and for any measurable partition $\{A_1, A_2, A_3, A_4\}$ of X satisfying (i)–(iv), $Q(f)$ is the Lebesgue measure of $A_1 \cup A_3$. A relation $\precsim$ on $\mathbf{D}$ may then be defined by stipulating that for all $f, g \in \mathbf{D}$,

$$f \precsim g \quad \text{iff} \quad Q(f) \le Q(g).$$

It is easy to verify that, with $\mathbf{N}$ defined by Definition 8.2, Axioms 1–8 are all satisfied. (Note that the consequences 0 and 1 satisfy Axiom 2, and all the other consequences belong to $Y \setminus Y^*$.) Also the countable additivity of Lebesgue measure ensures that Axiom 9 is satisfied. And for all $f \in \mathbf{D}$, if

$$f' = \left\{ \begin{array}{l} 1 \text{ on } A_1 \cup A_3 \\ 0 \text{ on } A_2 \cup A_4 \end{array} \right\}$$

then $f' \in \bigcup \mathbf{L}$ and $f' \sim f$; so Axiom 10 is satisfied. There are, however, no sets $\mathbf{E}$ and $\mathbf{E}'$ in $\mathbf{D}$ that satisfy Axiom 11. And Theorem D.12 fails to hold, as I will now show.

If there is a utility on Y, then there is in particular a utility u such that $u(0) = 0$ and $u(1) = 1$; furthermore, we may suppose that u is nonnegative. I will show that, for such a choice of u,

$$\int_{y_1=0}^{1} \int_{y_2=0}^{1} u(y_1, y_2)\, dy_2 dy_1 = 1 \qquad (\text{D.17})$$

and

$$\int_{y_2=0}^{1} \int_{y_1=0}^{1} u(y_1, y_2)\, dy_1 dy_2 = 0. \qquad (\text{D.18})$$

But by Tonelli's theorem (Royden 1968, p. 270), not both of (D.17) and (D.18) can be true. Hence there is no utility on Y.

To establish (D.17), observe that for all $y_1 \in [0, 1]$ there exists $f \in \mathbf{D}$ such that

$$f(x) = \left\{ \begin{array}{l} (y_1, x + y_1) \text{ for } x \in [-y_1, 1 - y_1] \\ 0 \text{ elsewhere} \end{array} \right\}.$$

Let p be the probability on X asserted to exist by Theorem B.1. It is easily verified that this probability is equal to one half of

289

Lebesgue measure on X. Hence

$$
\begin{aligned}
\mathcal{E}(u \circ f) &= \int_{x=-y_1}^{1-y_1} (u \circ f)\, dp \\
&= \int_{x=-y_1}^{1-y_1} u(y_1, x + y_1)\, dp \\
&= \frac{1}{2} \int_{y_2=0}^{1} u(y_1, y_2)\, dy_2.
\end{aligned}
\tag{D.19}
$$

Now if $f' \in \mathbf{D}$ is defined by the condition

$$
f' = \left\{
\begin{array}{l}
1 \text{ on } [-y_1, 1 - y_1] \\
0 \text{ elsewhere}
\end{array}
\right\},
$$

then $f' \sim f$. So since u is a utility on Y,

$$
\mathcal{E}(u \circ f) = \mathcal{E}(u \circ f') = p[-y_1, 1 - y_1] = \frac{1}{2}.
\tag{D.20}
$$

Combining (D.19) and (D.20) gives

$$
\int_{y_2=0}^{1} u(y_1, y_2)\, dy_2 = 1.
\tag{D.21}
$$

Integrating over y_1 in (D.21) yields (D.17).

The proof of (D.18) is similar. Observe that for all $y_2 \in [0, 1]$ there exists $g \in \mathbf{D}$ such that

$$
g(x) = \left\{
\begin{array}{l}
(y_2 - x, y_2) \text{ for } x \in [y_2 - 1, y_2] \\
0 \text{ elsewhere}
\end{array}
\right\}.
$$

Then

$$
\begin{aligned}
\mathcal{E}(u \circ g) &= \int_{x=y_2-1}^{y_2} (u \circ g)\, dp \\
&= \int_{x=y_2-1}^{y_2} u(y_2 - x, y_2)\, dp \\
&= \frac{1}{2} \int_{y_1=0}^{1} u(y_1, y_2)\, dy_1.
\end{aligned}
\tag{D.22}
$$

Now if $g' \in \mathbf{D}$ is the act with constant value 0, then $g' \sim g$. So since u is a utility on Y,

$$\mathcal{E}(u \circ g) = \mathcal{E}(u \circ g') = 0. \qquad \text{(D.23)}$$

Combining (D.22) and (D.23) gives

$$\int_{y_1=0}^{1} u(y_1, y_2) \, dy_1 = 0. \qquad \text{(D.24)}$$

Integrating over y_2 in (D.24) yields (D.18).

Bibliography

Allais, Maurice. 1979. "The So-Called Allais Paradox and Rational Decisions under Uncertainty." In Maurice Allais and Ole Hagen (eds.), *Expected Utility Hypotheses and the Allais Paradox*. Dordrecht: D. Reidel, pp. 437–681.

Anand, Paul. 1987. "Are the Preference Axioms Really Rational?" *Theory and Decision 23*:189–214.

Armendt, Brad. 1980. "Is There a Dutch Book Argument for Probability Kinematics?" *Philosophy of Science 47*:583–8.

Bacon, Francis. 1620. *Novum Organum*. English translation in James Spedding, Robert Leslie Ellis, and Douglas Denon Heath (eds.), *The Works of Francis Bacon*, vol. 8. Boston: Brown and Taggard, 1863.

Bar-Hillel, Maya, and Avishai Margalit. 1988. "How Vicious Are Cycles of Intransitive Choice?" *Theory and Decision 24*:119–45.

Bar-Hillel, Yehoshua, and Rudolf Carnap. 1953. "Semantic Information." *British Journal for the Philosophy of Science 4*:147–57.

Bell, David E. 1982. "Regret in Decision Making under Uncertainty." *Operations Research 30*:961–81.

Bernoulli, Daniel. 1738. "Specimen Theoriae de Mensura Sortis." *Commentarii Academiae Sicentarum Imperialis Petropolitanae V*:175–92. English translation by Louise Sommer in *Econometrica 22* (1954), 23–36; reprinted in Alfred N. Page (ed.), *Utility Theory: A Book of Readings*, New York: Wiley, 1968.

Blackburn, Simon. 1984. *Spreading the Word*. Oxford: Oxford University Press.

Blyth, Colin R. 1972. "Some Probability Paradoxes in Choice from among Random Alternatives." *Journal of the American Statistical Association* 67:366–73.

Bratman, Michael E. 1987. *Intention, Plans, and Practical Reason.* Cambridge, Mass.: Harvard University Press.

Broome, John. 1991. "Rationality and the Sure-Thing Principle." In Gay Meeks (ed.), *Thoughtful Economic Man.* Cambridge: Cambridge University Press, pp. 74–102.

Brown, Peter M. 1976. "Conditionalization and Expected Utility." *Philosophy of Science* 43:415–19.

Burks, Arthur W. 1977. *Chance, Cause, Reason.* Chicago: University of Chicago Press.

Campbell, Richmond, and Thomas Vinci. 1982. "Why Are Novel Predictions Important?" *Pacific Philosophical Quarterly* 63:111–21.

1983. "Novel Confirmation." *British Journal for the Philosophy of Science* 34:315–41.

Carnap, Rudolf. 1950. *Logical Foundations of Probability.* Chicago: Chicago University Press; 2d ed. 1962.

1952. *The Continuum of Inductive Methods.* Chicago: University of Chicago Press.

Cavendish, Henry. 1879. *The Electrical Researches of the Honorable Henry Cavendish.* Edited by J. Clerk Maxwell, Cambridge: Cambridge University Press.

Christensen, David. 1991. "Clever Bookies and Coherent Beliefs." *Philosophical Review* 100:229–47.

Cohen, L. Jonathan. 1977. *The Probable and the Provable.* Oxford: Oxford University Press.

Crick, Francis. 1988. *What Mad Pursuit.* New York: Basic Books.

Davidson, Donald. 1980. *Essays on Actions and Events.* New York: Oxford University Press.

1984. *Inquiries into Truth and Interpretation.* New York: Oxford University Press.

293

1985. "Incoherence and Irrationality." *Dialectica* 39:345–54.

Davidson, Donald, J. C. C. McKinsey, and Patrick Suppes. 1955. "Outlines of a Formal Theory of Value, I." *Philosophy of Science* 22:60–80.

de Finetti, Bruno. 1931. "Sul Significato soggettivo della probabilità." *Fundamenta Mathematicae* 17:298–329.

1937. "La Prévision: ses lois logiques, ses sources subjectives." *Annales de l'Institut Henri Poincaré* 7:1–68. Page references are to the English translation in Henry E. Kyburg Jr. and Howard E. Smokler (eds.), *Studies in Subjective Probability*, New York: Krieger, 1980.

Doppelt, Gerald. 1978. "Kuhn's Epistemological Relativism: An Interpretation and Defense." *Inquiry* 21:33–86.

Dorling, Jon. 1972. "Bayesianism and the Rationality of Scientific Inference." *British Journal for the Philosophy of Science* 23:181–90.

1974. "Henry Cavendish's Deduction of the Electrostatic Inverse Square Law from the Result of a Single Experiment." *Studies in History and Philosophy of Science* 4:327–48.

Duhem, Pierre. 1914. *La Théorie Physique: Son Objet, Sa Structure*. Paris: Marcel Rivière & Cie. 2d ed. Translated by P. P. Wiener as *The Aim and Structure of Physical Theory*, Princeton: Princeton University Press, 1954.

Eells, Ellery. 1982. *Rational Decision and Causality*. Cambridge: Cambridge University Press.

Ellis, Brian. 1988. "Solving the Problem of Induction Using a Values-Based Epistemology." *British Journal for the Philosophy of Science* 39:141–60.

Ellsberg, Daniel. 1961. "Risk, Ambiguity, and the Savage Axioms." *Quarterly Journal of Economics* 75:643–99. Reprinted in Peter Gärdenfors and Nils-Eric Sahlin (eds.), *Decision, Probability, and Utility*, Cambridge: Cambridge University Press, 1988.

Elster, Jon. 1983. *Sour Grapes*. Cambridge: Cambridge University Press.

Fishburn, Peter C. 1981. "Subjective Expected Utility: A Review of Normative Theories." *Theory and Decision 13:* 129–99.

Foley, Richard. 1987. *The Theory of Epistemic Rationality.* Cambridge, Mass.: Harvard University Press.

Franklin, Allan, and Colin Howson. 1985. "Newton and Kepler, A Bayesian Approach." *Studies in History and Philosophy of Science 16:*379–85. Reprinted in Allan Franklin, *The Neglect of Experiment,* Cambridge: Cambridge University Press, 1986, pp. 119–23.

Gärdenfors, Peter, and Nils-Eric Sahlin (eds.). 1988. *Decision, Probability, and Utility.* Cambridge: Cambridge University Press.

Gibbard, Allan. 1990. *Wise Choices, Apt Feelings.* Cambridge, Mass.: Harvard University Press.

Gibbard, Allan, and William L. Harper 1978. "Counterfactuals and Two Kinds of Expected Utility." In C. A. Hooker, J. J. Leach, and E. F. McClennen (eds.), *Foundations and Applications of Decision Theory,* vol. 1. Dordrecht: D. Reidel, pp. 123-62.

Giere, Ronald M. 1983. "Testing Theoretical Hypotheses." In John Earman (ed.), *Testing Scientific Theories.* Minneapolis: University of Minnesota Press, pp. 269–98.

Goldstein, Michael. 1983. "The Prevision of a Prevision." *Journal of the Americal Statistical Association 78:*817–19.

Good, I. J. 1967. "On the Principle of Total Evidence." *British Journal for the Philosophy of Science 17:*319–21. Reprinted in I. J. Good, *Good Thinking,* Minneapolis: University of Minnesota Press, 1983.

1969. "What Is the Use of a Distribution?" In P. R. Krishnaiah (ed.), *Multivariate Analysis–II.* New York and London: Academic Press, pp. 183–203.

1983. *Good Thinking.* Minneapolis: University of Minnesota Press.

1988. "The Interface between Statistics and Philosophy of Science." *Statistical Science 3*:386–412.

Goodman, Nelson. 1973. *Fact, Fiction, and Forecast.* 3d ed. Indianapolis: Hackett.

Gupta, Anil. 1989. "Remarks on Definitions and the Concept of Truth." *Proceedings of the Aristotelian Society 89*:227–46.

Halmos, Paul R. 1950. *Measure Theory.* New York: Van Nostrand.

Hammond, Peter J. 1988a. "Consequentialist Foundations for Expected Utility." *Theory and Decision 25*:25–78.

1988b. "Consequentialism and the Independence Axiom." In Bertrand R. Munier (ed.), *Risk, Decision and Rationality.* Dordrecht: D. Reidel, pp. 503–16.

Heilig, Klaus. 1978. "Carnap and de Finetti on Bets and the Probability of Singular Events: The Dutch Book Argument Reconsidered." *British Journal for the Philosophy of Science 29*:325–46.

Hempel, Carl G. 1960. "Inductive Inconsistencies." *Synthese 12*:439–69. Reprinted in Carl G. Hempel, *Aspects of Scientific Explanation*, New York: The Free Press, 1965.

1962. "Deductive-Nomological vs Statistical Explanation." In H. Feigl and G. Maxwell (eds.), *Minnesota Studies in the Philosophy of Science III.* Minneapolis: University of Minnesota Press, pp. 98–169.

Hempel, Carl G., and Paul Oppenheim. 1948. "Studies in the Logic of Explanation." *Philosophy of Science 15*:135–75. Reprinted in Carl G. Hempel, *Aspects of Scientific Explanation*, New York: The Free Press, 1965.

Horwich, Paul. 1982. *Probability and Evidence.* Cambridge: Cambridge University Press.

Howson, Colin, and Allan Franklin. 1991. "Maher, Mendeleev and Bayesianism." *Philosophy of Science 58*:574–85.

Howson, Colin, and Peter Urbach. 1989. *Scientific Reasoning.* La Salle, Ill.: Open Court.

Hughes, R. I. G. 1980. "Rationality and Intransitive Preferences." *Analysis* 40:132–4.

Hurley, S. L. 1989. *Natural Reasons.* New York: Oxford University Press.

Huygens, Christiaan. 1690. *Traité de la Lumière.* Leiden: Pierre vander Aa. Translated by S. P. Thompson as *Treatise on Light*, London: Macmillan, 1912.

Jeffrey, Richard C. 1965. *The Logic of Decision.* New York: McGraw-Hill; 2d ed., Chicago: University of Chicago Press, 1983.

1974. "Preference among Preferences." *Journal of Philosophy* 71:377–91. Reprinted in Richard C. Jeffrey, *The Logic of Decision*, 2d ed. Chicago: University of Chicago Press, 1983.

1983. "The Sure Thing Principle." In Peter D. Asquith and Thomas Nickles (eds.), *PSA 1982*, vol. 2. East Lansing: Philosophy of Science Association, pp. 719–30.

1987. "Risk and Human Rationality." *The Monist* 70:223–36.

1988. "Conditioning, Kinematics, and Exchangeability." In Brian Skyrms and William L. Harper (eds.), *Causation, Chance, and Credence*, vol. 1. Dordrecht: Kluwer, pp. 221–55.

Jeffreys, Harold. 1931. *Scientific Inference.* Cambridge: Cambridge University Press, 3d ed., 1973.

Jensen, Niels Erik. 1967. "An Introduction to Bernoullian Utility Theory I." *Swedish Journal of Economics* 69:163–83.

Kahneman, Daniel, Paul Slovic, and Amos Tversky (eds.). 1982. *Judgment under Uncertainty: Heuristics and Biases.* Cambridge: Cambridge University Press.

Kahneman, Daniel, and Amos Tversky. 1979. "Prospect Theory: An Analysis of Decisions under Risk." *Econometrica* 47:263–91. Reprinted in Peter Gärdenfors and Nils-Eric Sahlin (eds.), *Decision, Probability, and Utility*, Cambridge: Cambridge University Press, 1988.

Kaku, Michio, and Jennifer Trainer. 1987. *Beyond Einstein.* New York: Bantam.

Kaplan, Mark. 1981. "Rational Acceptance." *Philosophical Studies 40*:129-45.

Kashima, Yoshihisa, and Patrick Maher. 1992. "Framing of Decisions under Ambiguity." Department of Psychology, La Trobe University.

Kemeny, John G. 1955. "Two Measures of Simplicity." *Journal of Philosophy 52*:722-33.

Koertge, Noretta. 1979. "The Problem of Appraising Scientific Theories." In Peter D. Asquith and Henry E. Kyburg, Jr. (eds.), *Current Research in Philosophy of Science.* East Lansing: Philosophy of Science Association, pp. 228-51.

Kuhn, Thomas S. 1962. *The Structure of Scientific Revolutions.* Chicago: University of Chicago Press; 2d ed. 1970.

1977. "Objectivity, Value Judgment, and Theory Choice." In Thomas S. Kuhn, *The Essential Tension.* Chicago: University of Chicago Press, pp. 320-39.

1983. "Rationality and Theory Choice." *Journal of Philosophy 80*:563-70.

Kukla, Andre. 1991. "Criteria of Rationality and the Problem of Logical Sloth." *Philosophy of Science 58*:486-90.

Kyburg, Henry E. Jr. 1961. *Probability and the Logic of Rational Belief.* Middletown, Conn.: Wesleyan University Press.

1968. "Bets and Beliefs." *American Philosophical Quarterly 5*:63-78. Reprinted in Peter Gärdenfors and Nils-Eric Sahlin (eds.), *Decision, Probability, and Utility,* Cambridge: Cambridge University Press, 1988.

1978. "Subjective Probability: Criticisms, Reflections, and Problems." *Journal of Philosophical Logic 7*:157-80.

Lamport, Leslie. 1986. LaTeX: *A Document Preparation System.* Reading, Mass.: Addison-Wesley.

Laudan, Larry. 1977. *Progress and Its Problems.* Berkeley: University of California Press.

1981. "A Confutation of Convergent Realism." *Philosophy of Science 48*:19-49.

Lehman, R. Sherman. 1955. "On Confirmation and Rational Betting." *Journal of Symbolic Logic* 20:251–62.

Lehrer, Keith, and Carl Wagner. 1985. "Intransitive Indifference: The Semi-Order Problem." *Synthese* 65:249–56.

Leibniz, Gottfried Wilhelm. 1678. "Letter to Herman Conring." In Gottfried Wilhelm Leibniz, *Philosophical Papers and Letters*. Dordrecht: D. Reidel, pp. 186–91.

Levi, Isaac. 1967. *Gambling with Truth*. Cambridge, Mass.: MIT Press.

——— 1974. "On Indeterminate Probabilities." *Journal of Philosophy* 71:391–418.

——— 1976. "Acceptance Revisited." In Radu J. Bogdan (ed.), *Local Induction*. Dordrecht: D. Reidel, pp. 1-71.

——— 1980. *The Enterprise of Knowledge*. Cambridge, Mass.: MIT Press.

——— 1984. *Decisions and Revisions*. Cambridge: Cambridge University Press.

——— 1986. *Hard Choices*. Cambridge: Cambridge University Press.

——— 1987. "The Demons of Decision." *The Monist* 70:193–211.

Lewis, David. 1981. "Causal Decision Theory." *Australasian Journal of Philosophy* 59:5–30.

Luce, R. Duncan. 1956. "Semiorders and a Theory of Utility Discrimination." *Econometrica* 24:178–91.

McClennen, Edward F. 1983. "Sure Thing Doubts." In Bernt P. Stigum and Fred Wenstøp (eds.), *Foundations of Utility and Risk Theory with Applications*. Dordrecht: D. Reidel, pp. 117–36. Reprinted in Peter Gärdenfors and Nils-Eric Sahlin (eds.), *Decision, Probability, and Utility*, Cambridge: Cambridge University Press, 1988.

——— 1990. *Rationality and Dynamic Choice*. Cambridge: Cambridge University Press.

MacCrimmon, Kenneth R. 1968. "Descriptive and Normative Implications of Decision Theory." In Karl Borch and

Jan Mossin (eds.), *Risk and Uncertainty*. New York: St. Martin's Press, pp. 3–23.

MacCrimmon, Kenneth R., and Stig Larsson. 1979. "Utility Theory: Axioms versus 'Paradoxes'." In Maurice Allais and Ole Hagen (eds.), *Expected Utility Hypotheses and the Allais Paradox*. Dordrecht: D. Reidel, pp. 333–409.

Maher, Patrick. 1984. Rationality and Belief. Doctoral dissertation, University of Pittsburgh.

1986a. "What Is Wrong with Strict Bayesianism?" *PSA 1986*, vol. 1:450–7.

1986b. "The Irrelevance of Belief to Rational Action." *Erkenntnis 24*:363–84.

1987. "Causality in the Logic of Decision." *Theory and Decision 22*:155–72.

1988. "Prediction, Accommodation, and the Logic of Discovery." *PSA 1988*, vol. 1:273–85.

1989. "Levi on the Allais and Ellsberg Problems." *Economics and Philosophy 5*:69–78.

1990a. "How Prediction Enhances Confirmation." In J. Michael Dunn and Anil Gupta (eds.), *Truth or Consequences: Essays in Honor of Nuel Belnap*. Dordrecht: Kluwer, pp. 327–43.

1990b. "Symptomatic Acts and the Value of Evidence in Causal Decision Theory." *Philosophy of Science 57*:479–98.

1990c. "Why Scientists Gather Evidence." *British Journal for the Philosophy of Science 41*:103–19.

1992. "Diachronic Rationality." *Philosophy of Science 59*:120–41.

In press. "Howson and Franklin on Prediction." *Philosophy of Science*.

Maher, Patrick, and Yoshihisa Kashima. 1991. "On the Descriptive Adequacy of Levi's Decision Theory." *Economics and Philosophy 7*:93–100.

300

Mao Tse-tung. 1966. *Quotations from Chairman Mao Tse-Tung.* Peking: Foreign Languages Press.

Markowitz, Harry M. 1959. *Portfolio Selection.* New York: Wiley.

Miller, David. 1974. "Popper's Qualitative Theory of Verisimilitude." *British Journal for the Philosophy of Science* 25:166–77.

Moore, G. E. 1942. "A Reply to My Critics." In Paul Arthur Schilpp (ed.), *The Philosophy of G. E. Moore.* Evanston, Ill.: Northwestern University, pp. 535–677.

Morrison, Donald. 1967. "On the Consistency of Preferences in Allais' paradox." *Behavioral Science* 5:225–42.

Moscowitz, Herbert. 1974. "Effects of Problem Representation and Feedback on Rational Behavior in Allais and Morlat-type Problems." *Decision Sciences* 5:225–41.

Newton, Isaac. 1726. *Philosophiæ Naturalis Principia Mathematica.* 3d ed. English translation by Andrew Motte, revised by Florian Cajori, Berkeley: University of California Press, 1934.

Newton-Smith, W. H. 1981. *The Rationality of Science.* Boston: Routledge and Kegan Paul.

Niiniluoto, Ilkka. 1984. *Is Science Progressive?* Dordrecht: D. Reidel.

1986. "Truthlikeness and Bayesian Estimation." *Synthese* 67:321–46.

1987. *Truthlikeness.* Dordrecht: D. Reidel.

Oddie, Graham. 1981. "Verisimilitude Reviewed." *British Journal for the Philosophy of Science* 32:237–65.

Packard, Dennis J. 1982. "Cyclical Preference Logic." *Theory and Decision* 14:415–26.

Peirce, Charles S. 1883. "A Theory of Probable Inference." In Charles S. Peirce (ed.), *Studies in Logic.* Boston: Little, Brown, pp. 126–81.

Popper, Karl R. 1959. *The Logic of Scientific Discovery.* New York: Harper and Row; 2d ed., 1968.

1963. *Conjectures and Refutations.* New York: Harper and Row; 2d ed., 1965.

1972. *Objective Knowledge.* Oxford: Oxford University Press; rev. ed., 1979.

Raiffa, Howard. 1968. *Decision Analysis.* Reading, Mass.: Addison-Wesley.

Railton, Peter. 1984. "Alienation, Consequentialism, and the Demands of Morality." *Philosophy & Public Affairs* *13*:134–71.

Ramsey, F. P. 1926. "Truth and Probability." In F. P. Ramsey, *Foundations.* Atlantic Highlands, N.J.: Humanities Press, pp. 58–100.

Rawls, John. 1971. *A Theory of Justice.* Cambridge, Mass.: Harvard University Press.

Rosenkrantz, Roger D. 1977. *Inference, Method and Decision.* Dordrecht: D. Reidel.

1981. *Foundations and Applications of Inductive Probability.* Atascadero, Calif.: Ridgeview.

Royden, H. L. 1968. *Real Analysis.* 2d ed. New York: Macmillan.

Salmon, Wesley C. 1967. *The Foundations of Scientific Inference.* Pittsburgh: University of Pittsburgh Press.

Savage, Leonard J. 1954. *The Foundations of Statistics.* New York: John Wiley; 2d ed., New York: Dover, 1972.

1971. "The Elicitation of Personal Probabilities and Expectations." *Journal of the American Statistical Association* *66*:783–801.

Schervish, Mark J., Teddy Seidenfeld, and Joseph B. Kadane. 1984. "The Extent of Non-Conglomerability of Finitely Additive Probabilities." *Zeitschrift für Wahrscheinlichkeitstheorie und verwandte Gebiete 66*:205–26.

Schick, Frederic. 1986. "Dutch Bookies and Money Pumps." *Journal of Philosophy 83*:112–19.

Seidenfeld, Teddy. 1988. "Decision Theory Without 'Independence' or without 'Ordering,' What Is the Difference?" *Economics and Philosophy* 4:267–90.

Seidenfeld, Teddy, and Mark Schervish. 1983. "A Conflict between Finite Additivity and Avoiding Dutch Book." *Philosophy of Science* 50:398–412.

Sen, Amartya K. 1971. "Choice Functions and Revealed Preference." *Review of Economic Studies* 38:307–17.

Shannon, C. E., and W. Weaver. 1949. *The Mathematical Theory of Communication.* Urbana: University of Illinois Press.

Shimony, Abner. 1955. "Coherence and the Axioms of Confirmation." *Journal of Symbolic Logic* 20:1–28.

——— 1970. "Scientific Inference." In Robert Colodny (ed.), *The Nature and Function of Scientific Theories.* Pittsburgh: University of Pittsburgh Press, pp. 79–172.

Skyrms, Brian. 1984. *Pragmatics and Empiricism.* New Haven, Conn.: Yale University Press.

——— 1987a. "Coherence." In Nicholas Rescher (ed.), *Scientific Inquiry in Philosophical Perspective.* Lanham, Md.: University Press of America, pp. 225–41.

——— 1987b. "Dynamic Coherence and Probability Kinematics." *Philosophy of Science* 54:1–20.

——— 1990a. "The Value of Knowledge." In C. Wade Savage (ed.), *Scientific Theories.* Minneapolis: University of Minnesota Press, pp. 245–66.

——— 1990b. *The Dynamics of Rational Deliberation.* Cambridge, Mass.: Harvard University Press.

——— In press. "A Mistake in Dynamic Coherence Arguments?" *Philosophy of Science.*

Slovic, Paul, and Amos Tversky. 1974. "Who Accepts Savage's Axiom?" *Behavioral Science* 19:368–73.

Sober, Elliott. 1975. *Simplicity.* London: Oxford University Press.

Stich, Stephen P. 1983. *From Folk Psychology to Cognitive Science*. Cambridge, Mass.: MIT Press.

Szaniawski, Klemens. 1976. "Types of Information and Their Role in the Methodology of Science." In Marian Przełęcki, Klemens Szaniawski, and Ryszard Wójcicki (eds.), *Formal Methods in the Methodology of Empirical Sciences*. Dordrecht: D. Reidel, pp. 297–308.

Talbott, W. J. 1991. "Two Principles of Bayesian Epistemology." *Philosophical Studies 62*:135–50.

Taylor, Charles. 1982. "Rationality." In Martin Hollis and Steven Lukes (eds.), *Rationality and Relativism*. Cambridge, Mass.: MIT Press, pp. 87–105.

Teller, Paul. 1973. "Conditionalization and Observation." *Synthese 26*:218–58.

1976. "Conditionalization, Observation, and Change of Preference." In W. Harper and C. Hooker (eds.), *Foundations of Probability Theory, Statistical Inference, and Statistical Theories of Science*. Dordrecht: D. Reidel, pp. 205–53.

Thagard, Paul R. 1988. *Computational Philosophy of Science*. Cambridge, Mass.: MIT Press.

Tichý, Pavel. 1974. "On Popper's Definitions of Verisimilitude." *British Journal for the Philosophy of Science 25*:155–60.

1978. "Verisimilitude Revisited." *Synthese 38*:175–96.

Tversky, Amos. 1969. "Intransitivity of Preferences." *Psychological Review 76*:31–48.

Urbach, Peter. 1983. "Intimations of Similarity: The Shaky Basis of Verisimilitude." *British Journal for the Philosophy of Science 34*:266–75.

van Fraassen, Bas C. 1980. *The Scientific Image*. New York: Oxford University Press.

1983. "Glymour on Evidence and Explanation." In John Earman (ed.), *Testing Scientific Theories*. Minneapolis: University of Minnesota Press, pp. 165–76.

1984. "Belief and the Will." *Journal of Philosophy 81*:235–56.

1985. "Empiricism in the Philosophy of Science." In Paul M. Churchland and Clifford A. Hooker (eds.), *Images of Science*. Chicago: University of Chicago Press, pp. 245–308.

1989. *Laws and Symmetries*. New York: Oxford University Press.

In press. "Rationality Does Not Require Conditionalization." In Edna Ullmann-Margalit (ed.), *The Israel Colloquium Studies in the History, Philosophy, and Sociology of Science*, vol. V. Dordrecht: Kluwer.

von Neumann, John, and Oskar Morgenstern. 1947. *Theory of Games and Economic Behavior*. 2d ed. Princeton, N.J.: Princeton University Press.

Wakker, Peter. 1988. "Nonexpected Utility as Aversion of Information." *Journal of Behavioral Decision Making* 1:169–75.

Whewell, William. 1847. *The Philosophy of the Inductive Sciences*. 2d ed. London: Parker.

Wilson, Timothy D., Jay G. Hull, and Jim Johnson. 1981. "Awareness and Self-Perception: Verbal Reports on Internal States." *Journal of Personality and Social Psychology* 40:53–71.

Worrall, John. 1989. "Fresnel, Poisson and the White Spot: The Role of Successful Predictions in the Acceptance of Scientific Theories." In David Gooding, Trevor Pinch, and Simon Schaffer (eds.), *The Uses of Experiment*. Cambridge: Cambridge University Press, pp. 135–57.

Index

probability kinematics, 127–8
properties α and β, 45–7
proposition, 133, 186

qualified Bayesianism, 29–33

Raiffa, Howard, 36, 65, 70–1, 74
Railton, Peter, 8n
Ramsey, F. P., 9, 91–2, 152, 221
rationality, 23–9, 61
Rawls, John, 87
realism, scientific, 240–4
Reflection, 105–6; argument for,
  106–7, 110–13; and conditionalization, 121; counterexamples to,
  107–10; true status of, 116–20
representation theorem, 9–12, 20–1,
  30, 195–7, 206–7
representor, 21
Rescher, Nicholas, xi
rigidity, 38–41, 75–6, 78, 81
Rosen, Gideon, 241n
Rosenkrantz, Roger D., 166–8, 176,
  215

Salmon, Wesley C., xi, 29, 91–2
Savage, Leonard J., 9, 10, 11, 13, 19n,
  63, 76–8, 179, 182–5, 196n, 249, 250,
  270
Schervish, Mark J., 196n, 267
Schick, Frederic, 98
Schmitt, Frederick, xii
Schwartz, John, 215
scientific realism, 240–4
scientific values, 209–10
Seidenfeld, Teddy, xi, 81, 196n, 267,
  272n
Sen, Amartya, 23, 45–7
Shannon, C. E., 234
shifting focus, 49–51
Shimony, Abner, 88n, 97
simplicity, 215–16
Skyrms, Brian, 20, 68, 102–4, 106,
  110, 115n
Slovic, Paul, 64–8, 87
Sober, Elliott, 215
state, 1–5, 186
Stich, Stephen P., 153

strict preference, 14
Suppes, Patrick, 36
sup(remum), 143n
sure-thing principle, 10n, 76–9
SVET, definition of, 173; see also
  evidence, value of
synchronic separability, 73
Szaniawski, Klemens, 234

Talbott, W. J., 113
Taylor, Charles, 24
Teller, Paul, 120, 123–4
Thagard, Paul, 88–9
Tichý, Pavel, 213, 221–4, 226
transitivity, 10, 11, 12, 14, 20;
  arguments for, 36–47; axiom on,
  190; objections to, 47–60; and
  popular endorsement, 34–6
truth, 208–9; distance from, 237–40;
  and impartiality, 214; respect for,
  210–13
Tversky, Amos, 35n, 64–8, 71–2, 87

Urbach, Peter, 86, 94, 224
utility, 9, 88; expected, 1–5; scientific,
  210

van Fraassen, Bas, 21n, 93, 105, 106,
  113, 125–6, 158–61, 165n, 221n,
  240–3
verisimilitude: defined, 228; expected,
  229–30; need for, 141–2, 218–20;
  Popper's theory of, 220–2;
  subjective approach to, 224–7;
  Tichý's theory of, 222–4
Vinci, Thomas, 86n
von Neumann, John, 97, 254

Wagner, Carl, 38n
Wagner, Steven, xii
Wakker, Peter, 79–81
weak order, 10
Whewell, William, 86
will, weakness of, 17–18, 30–1
Wilson, Timothy D., 154
Worrall, John, 160n
Wu, Wei-ming, xii

309